PHOTO ATLAS OF MINERAL PSEUDOMORPHISM

PHOTO ATLAS OF MINERAL PSEUDOMORPHISM

J. THEO KLOPROGGE

School of Earth and Environmental Sciences, University of Queensland
St. Lucia, Australia and Department of Chemistry, College of Arts and Sciences
University of the Philippines Visayas, Miagao, Philippines

ROBERT LAVINSKY

The Arkenstone Ltd (iRocks.com)- Dallas, TX, USA

ELSEVIER

Elsevier
Radarweg 29, PO Box 211, 1000 AE Amsterdam, Netherlands
The Boulevard, Langford Lane, Kidlington, Oxford OX5 1GB, United Kingdom
50 Hampshire Street, 5th Floor, Cambridge, MA 02139, United States

Notices
Knowledge and best practice in this field are constantly changing. As new research and experience broaden our understanding, changes in research methods, professional practices, or medical treatment may become necessary.

Practitioners and researchers must always rely on their own experience and knowledge in evaluating and using any information, methods, compounds, or experiments described herein. In using such information or methods they should be mindful of their own safety and the safety of others, including parties for whom they have a professional responsibility.

To the fullest extent of the law, neither the Publisher nor the authors, contributors, or editors, assume any liability for any injury and/or damage to persons or property as a matter of products liability, negligence or otherwise, or from any use or operation of any methods, products, instructions, or ideas contained in the material herein.

Library of Congress Cataloging-in-Publication Data
A catalog record for this book is available from the Library of Congress

British Library Cataloguing-in-Publication Data
A catalogue record for this book is available from the British Library

ISBN: 978-0-12-803674-7

For information on all Elsevier publications visit our
website at https://www.elsevier.com/books-and-journals

Working together
to grow libraries in
developing countries

www.elsevier.com • www.bookaid.org

Publisher: Candice Janco
Acquisition Editor: Amy Shapiro
Editorial Project Manager: Tasha Frank
Production Project Manager: Paul Prasad Chandramohan
Designer: Mathew Limbert

Typeset by TNQ Books and Journals

Contents

1

Introduction

INTRODUCTION

The phenomenon of pseudomorphism in mineralogy was recognized more than two centuries ago and over the years many examples have been collected by various museums around the world. If one would like to give a simple definition of a pseudomorph, one could say that a pseudomorph is a mineral that has the outward form proper to another species of mineral whose place it has taken through the action of some agency (Frondel, 1935). This definition is very strict and excludes a number of replacements that are considered pseudomorphs these days, such as regular cavities left by the removal of a crystal from its matrix (molds) because these are voids and not solids, and also excludes those cases in which the organic material has been replaced by some other mineral because the original substance here is not a mineral. This means that it also includes any mineral change in which the outlines of the original mineral are preserved, whether this surface is a euhedral crystal form or the irregular bounding surface of an embedded grain or of an aggregate, or any mineral change that has been accomplished without change of volume, as evidenced by the undistorted preservation of an original texture or structure, whether this be the equal volume replacement of a single crystal or of a rock mass on a geologic scale. Based on these definitions pseudomorphs occur quite regularly but are not always recognized as such by geologists and their importance for gaining a better understanding of the mineralogical and petrological history.

HISTORICAL PERSPECTIVE

The term pseudomorph (French = la pseudomorphose) was first used by René Just Haüy (1743–1822) in his ground-breaking work *Traité de Mineralogie* (Haüy, 1801) (Fig. 1.1). Haüy did not, however, use it in its present sense. As part of a general discussion about types of concretions he used the term based on a combination of the Greek words pseudo (false) and morph (form) for mineral bodies that owed their outward form to circumstances other than their own unique powers of crystallization or formation. Most of the discussion was focused on fossils, especially fossil shells and petrified wood. He also included at the end of the discussion, the comment as translated from French:

...bodies which have a false and deceitful figure

which

...present in a very remarkable manner foreign or strange forms which they have in some measure obtained from other bodies which had received them from nature.

FIGURE 1.1 René Just Haüy by Ambrose Tardieu (1788–1841).

After considerable discussion of what now would be classified as fossils, he added that:

The mineral kingdom also has its pseudomorphoses. We find some substances of this kingdom under crystalline forms which are only borrowed; and it is probable that, in some cases at least, the new substance has been substituted gradually for that which has ceded its place to it as we suppose takes place with respect to petrified wood.

Photo Atlas of Mineral Pseudomorphism
http://dx.doi.org/10.1016/B978-0-12-803674-7.00001-3

FIGURE 1.2 Johann Reinhard Blum (© Universitätsbibliothek Heidelberg).

He ended up by giving a definition. After defining stalactites and incrustations as the two other forms of concretion, Haüy wrote that:

> the pseudomorphoses is a concretion endowed with a form foreign to its substance and for which it is indebted to its molecules filling a space formerly occupied by a body of the same form.

The earliest reference to what would now be called a pseudomorph was apparently in a privately printed, very rare descriptive mineralogy text by Franz Joseph Anton Estner (1739–1803?), a German born Abbot and mineralogist resident in Vienna. It was "…designed both for the beginner and the lover of minerals" (Schuh, 2000). This great, three-volume rarity was privately published between 1794 and 1804 in Vienna by Estner under the title *Versuch einer Mineralogie für Anfänger und Liebhaber nach des Herrn Bergcommissionsraths Werner's Methode* (Estner, 1794–1804). Estner was a student of and great admirer of Werner.

The extended and exhaustive works of Johann Reinhard Blum (1802–83) (Fig. 1.2), *Die Pseudomorphosen des Mineralreichs* (Stuttgart) appeared in 1843 (Blum, 1843). He issued supplements in 1847, 1852, 1863, and 1879 (Blum, 1847, 1852, 1863, 1879). According to J.R. Blum (1843), Estner in Vol. 1, 143–145 (1794–1804) used the term After-Kristalle. Estner named crystals that had a different form than what their composition would dictate as "After-

NACHTRAG

ZU DEN

PSEUDOMORPHOSEN

DES

MINERALREICHS

NEBST EINEM ANHANGE

ÜBER DIE

VERSTEINERUNGS- UND VERERZUNGS-MITTEL ORGANISCHER KÖRPER

VON

Dr. J. REINHARD BLUM,

ausserordentlicher Professor an der Universität zu Heidelberg und mehrerer gelehrten Gesellschaften Mitgliede.

STUTTGART.

E. SCHWEIZERBART'SCHE VERLAGSHANDLUNG.

1847.

Kristalle" (Johs, 1981). The term After-Kristalle apparently was used in the present sense of pseudomorph by Werner (1749–1817) and taken up by his students, which in that era included a good percentage of the influential mineralogists in Europe. After-Kristalle obviously has precedence, but by the second half of the nineteenth century it had been replaced almost completely by pseudomorph. Werner's usage is often cited to 1811 because that is when Carl August Siegfried Hoffmann (1760–1813) started publishing his monumental four volumes *Handbuch der Mineralogie* (Hoffmann, 1811–1817). Hoffmann was a keen student and Werner was his mentor. Hoffmann "…accepts all of Werner's theories concerning mineral formation and classification" (Schuh, 2000). It seems; therefore, fair to attribute the term After-Kristalle to Werner. Werner had divided crystals into two categories:

1. Wesentliche (essential or true)
2. After-Kristalle

Johann August Breithaupt (1791–1873) was another famous student of Werner. He later had a 40-year career as a full professor of mineralogy at his alma mater, the mining academy at Freiberg. In 1815, he published *Über die Aechtheit der Kristalle* (*Concerning the genuineness of crystals*). This is the first monograph devoted to pseudomorphs (Breithaupt, 1815).

George Amadeus Carl Friedrich Naumann (1797–1873), an important pioneer European mineralogist and crystallographer, defined a pseudomorph (Naumann, 1846) as

> crystalline or amorphous body that without itself being a crystal shows the crystal form of another mineral [so nennt man nämlich diejenigen krystallinischen oder amorphen Mineralkörper, welche ohne selbst Krystalle zu sein, die Krystallform eines anderen Minerals zeigen].

That Naumann favored a more restrictive definition than that advocated by some of his colleagues is shown by his remark.

> The crystal forms of pseudomorphs are usually quite well preserved [erhalten] and easily recognized with sharp, well-formed faces [Diese Krystallformen der Pseudomorphosen sind meist sehr wohl erhalten und leicht erkennbar, ja zuweilen ganz sharfkantig und glatt].

In the same work, he distinguished (1) Umhüllung-Pseudomorphosen, (2) Ausfüllungs-Pseudomorphosen, and (3) Metasomatische Pseudomorphosen. Umhüllungs-Pseudomorphosen: literally wrapping around, enveloping, or encasing type pseudomorphs. It usually refers to the case where a crystal is encrusted by another mineral and then later the first crystal is dissolved away or partially dissolved away. This type of pseudomorph forms commonly as quartz after calcite. Often called perimorph (the new mineral is on the periphery of the first one). Ausfüllungs-Pseudomorphosen: literally filling in or filling up type pseudomorphs. Metasomatish-Pseudomorphosen (metasomatic pseudomorph). Metasomatism is a term primarily applied to ore deposits.

> It is the process of practically simultaneous capillary solution and deposition by which a new mineral of partly or wholly different composition may grow in the body of an old mineral or mineral aggregate. *Bates, R.L., Jackson, J.A., 1980. Glossary of Geology. American Geological Institute, Falls Church, VA.*

In common with Johann Reinhard Blum (1802–83), the greatest nineteenth century expert on pseudomorphs (Blum, 1843, 1847, 1852, 1863, 1879), and other European authorities, Naumann (1846) further divided the first category into Umhüllings, Ausfüllings, and Verdrängungs (replacement) pseudomorphs. Naumann and George Landgrebe (1802–72) a German chemist and mineralogist who wrote a very influential book on pseudomorphs (Landgrebe, 1841) divided the second category metasomatische into the following classes of pseudomorphs:

1. Formed by molding (Abformung)
   a. Molding by being coated; e.g., quartz encrusting calcite that then disappears.
   b. By having the new mineral totally replace the original one (Abforming durch Ausfüllung); e.g., the famous talc pseudomorphs after quartz crystals from Göpfersgrun, near Wunsiedel in Bavaria.
2. Formed by alteration (durch Umwandlung).
   a. Alteration without gain or loss of new components (Umwandlung ohne Abgabe oder Aufnahme von Stoffen); e.g., aragonite altered to calcite.
   b. Alteration with the loss of a component (Umwandlung mit Verlust von Bestandteilen); e.g., laumontite loss of water turning it into a crumbling powder.

c. Alteration with the addition of components (Umwandlung mit Aufname von Bestandteilen); e.g., anhydrite altered to gypsum.

d. Alteration with exchange of components (Umwandlung mit Austausch von Stoffen); e.g., feldspar altered to cassiterite.

In 1855, Gustav Georg Winkler (1820—96) published a small monograph that summarizes the works of Blum, Haidinger, Landgrebe, and others and contains a large descriptive section (Winkler, 1855).

It is clear from the earlier section, the German language's penchant for sticking a number of unrelated words together to make a new definition makes the original literature of pseudomorphs difficult to read. One should remember that in the early days mineralogists writing in English generally had no choice but to try to translate pseudomorph terms from German into English.

Many other definitions, descriptions, and classifications of pseudomorphs have been offered over the years. The vast majority are written in German, a lesser number in French, and very few in English. Those written in English were primarily derived from German authors because during the nineteenth century particularly, virtually all acknowledged authorities on pseudomorphs wrote in German. Thus the roots to the English terms applicable to pseudomorphs are derived from German language terms. The most notable exception is the term pseudomorph itself, which is derived from Greek, but first appeared in a French publication.

The earliest systematic treatment of pseudomorphs in English was made by Wilhelm Karl von Haidinger in *Brewster's Edinburgh Journal of Science* (Haidinger, 1827, 1828, 1829) (Fig. 1.3). Around 1827 Haidinger's chief interest turned to pseudomorphs because they are one of the few unambiguous indicators of a mineral change that must have taken place in the past. Following Humphry Davy (1778—1829), he attributed the motion and replacement of particles to electrochemical forces, drawing a parallel between this interchange of constituents and the behavior of a solution in electrolytic dissociation. The calcite—dolomite transformation led Haidinger to the premise that percolating saline solutions (Gebirgsfeuchtigkeit) containing $MgSO_4$ replace $2CaCO_3$ by $CaCO_3 + MgCO_3$ molecules. This process was thought to proceed at elevated pressure and temperature, with precipitation of gypsum. Most of the gypsum would be taken into solution and redeposited at greater depths. Under surface conditions, the reverse reaction occurs, that is, dedolomitization, with lime replacing magnesium.

Ueber die

PSEUDOMORPHOSEN

im

MINERALREICHE

und

verwandte Erscheinungen

von

Dr. Georg Landgrebe,

Mitgliede mehrerer gelehrten Gesellschaften.

CASSEL,

Verlag von J. J. Bohné.

1841.

FIGURE 1.3 Wilhelm Karl von Haidinger 1844 lithography by Josef Kriehuber (1800—76).

Perhaps the best modern classification of pseudomorphs was written by Hugo Strunz in the German semiprofessional mineralogy and geology journal *Der Aufschluss* (Strunz, 1982). The classification presented here is based on that of Strunz with a few additions. It is more inclusive than what is generally encountered. This has the advantage that it includes areas and modern concepts that have attracted the attention of modern mineralogists. The collector, of course, is always free to keep or discard whatever parts he or she chooses. Strunz recognized four basic types of pseudomorphs, some of which were not even thought of in earlier works.

1. Paramorphs, sometimes called transformation pseudomorphs; e.g., acanthite after argentite. He makes metamict a subcategory.
2. Exsolution pseudomorphs (Entmischungs-Pseudomorph); e.g., magnetite and rutile after ilmenite.
3. Replacement (Verdrängungs) pseudomorphs. Here he has four subcategories.
   a. giving up a component; e.g., malachite after azurite
   b. taking up a component; e.g., talc after quartz
   c. replacement (austausch) of a component; e.g., limonite after pyrite or fluorite after calcite
   d. exchange of all components; e.g., native copper after aragonite
4. Perimorphs; e.g., quartz after calcite.

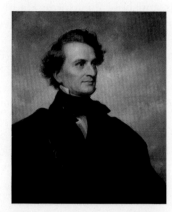

FIGURE 1.4 James Dwight Dana, 1858 oil on canvas by Daniel Huntington (1816–1906), Yale Art Gallery, Yale University, New Haven, CT.

What is perhaps the most authoritative and recent work on pseudomorphs published in the English language was written by Clifford Frondel of Harvard University in 1935. It was based on an extensive study of pseudomorphs in the collections of the American Museum of Natural History in New York as well as a great familiarity with European works on pseudomorphs (Frondel, 1935). Perhaps the most useful recent definition is offered by Sinkankas (1964):

> If a crystal changes chemically or structurally, yet keeps the shape of the original, it is called a pseudomorph or 'false form'; it looks like a crystal of one species but is composed of another.

For American collectors, probably the most influential of all classifications is that first proposed by James D. Dana (1813–95) in the middle of the last century (Dana, 1845) (Fig. 1.4). For more than a century and a half, Americans have been particularly influenced by the first major scientific work published by a native-born American: *A System of Mineralogy* by James Dwight Dana (1837). In the first edition, he gives a very brief treatment of pseudomorphs, but defines a pseudomorph as

> A pseudomorphous crystal, is one which possesses a form that is foreign to it, which it has received from some other cause, distinct from its own powers of crystallization.

He also makes an observation that may not be appreciated by all pseudomorphs collectors:

> Pseudomorphs crystals are distinguished, generally by their rounded angles, dull surfaces, destitution of cleavage joints and often granular composition. The surfaces are frequently drusy or covered with minute crystals. Occasionally, however, the resemblance to real [sic] crystals is so perfect, that they are distinguished with difficulty.

The treatment remained essentially unchanged in the second edition (Dana, 1844). In the third edition (Dana, 1850), he summarized the previous descriptive work on pseudomorphs and compiled a list of the 82 pseudomorphs reported to that time. This work seems to be the first important treatment of pseudomorphs by an American author. Dana's work has had great influence on how American collectors and writers conceptualize pseudomorphs. Note that the categories are somewhat different and less inclusive than Dr. Strunz's. In the fourth edition (Dana, 1854), Dana formalized as follows:

1. Pseudomorphs by alteration: Those that formed by the gradual change of composition in a species; e.g., change of augite to steatite, or azurite to malachite.

2. Pseudomorphs by substitution: Those that formed by the replacement of a mineral that has been removed or is gradually undergoing removal; e.g., petrifaction of wood.

3. Pseudomorphs by encrustation: Those formed through the encrustation of a crystal that may have subsequently dissolved away; often the cavity afterward is filled (or partially filled) by infiltration; e.g., change of fluorite to quartz.

4. Pseudomorphs by paramorphism: Those formed when a mineral passes from one (dimorphous) state to another; e.g., change of aragonite to calcite or beta quartz to alpha quartz.

5. Perimorph: Not to be confused with paramorph. A perimorph is a special type of pseudomorph which is formed when one mineral is encrusted by another, and then the original mineral is leached out leaving a hollow shell in the form of the original mineral. Epimorphs are a special case of a perimorph. L.P. Gratacap in his *Popular Guide to Minerals* (Gratacap, 1912) defines an epimorph as "a pseudomorph formed by encrustation as when quartz coats calcite, concealing the covered mineral completely though assuming the crystalline form of the calcite. Such phases of pseudomorphism are called epimorphs."

Some authors specify that an epimorph is a thin coating that occurs when a pyrite crystal is altered on the surface to limonite. Some authorities quite rightly object to the term perimorph because it phonetically is too close to paramorph and the whole subject of pseudomorphs is confusing enough as it is.

Dana in the fourth edition (1854) has a detailed discussion of pseudomorphs and how they form. He gives a classification and an exhaustive (86!) list of pseudomorphs known at that time. For some reason this valuable section was dropped from later editions.

A pseudomorph is described as being "after" the mineral whose outer form it has; e.g., quartz after fluorite (Bates and Jackson, 1980). This convention was introduced by J.R. Blum (1843, 1847, 1852, 1863, 1879), who assembled what is probably the world's most extensive pseudomorph collection in the world. It is now preserved in the basement of the Geology building at Yale. It is for the most part not very pretty to look at but it is a very important historical and scientific collection.

The most extensive modern work on pseudomorphs, from a collector's point of view, is the November, 1981, issue of the German mineral collectors' magazine *Lapis*. The entire issue was devoted to pseudomorphs with articles by various European authorities. This is now a sought after collector's item. In addition to it, the same publisher also published in 2012 the title in German, ExtraLapis no. 43 Pseudomorphosen.

Classification

A classificatory system can be very helpful not just for organizing a collection of specimens but for appreciating relationships among them. Unfortunately, there is no single, well agreed upon, classificatory scheme for pseudomorphs although the one by Dana has been particularly influential among American authors. The best modern discussion and classification of pseudomorphs is a long article by Hugo Strunz (1982). The classification scheme presented here is based on his work.

Below is an outline of nine categories into which pseudomorphs might be divided. These categories are basically those of Strunz, with a few additions. Included are, in the interest of comprehensiveness, certain mineralogical phenomena which are included in only a few of the many classifications of pseudomorphs published over the years. In any classificatory attempt there is a natural tension between "lumpers" and "splitters." Some collectors and some scientists may be inclined to broader and others to more restricted categories. The categories below cover as broad a range as possible, working under the premise that the curator, collector, or scientist can always discard any that he or she feels do not belong or do not fit their purposes. The categories have been arranged from simple to more complex.

1. Paramorphs. Paramorphs are the result of polymorphism. Polymorphism is the existence of a chemical compound or element in two or more crystal structures. Carbon as diamond or graphite is a familiar example. A mineral material called cliftonite found in certain meteorites is graphite pseudomorphing diamond, alpha (low temperature) quartz pseudomorphs (paramorphs) after beta (high temperature) quartz are well known.

   If two forms are possible, they are dimorphs. If there are three forms, they are trimorphs and so on. Rutile, anatase, and brookite are trimorphs of titanium dioxide TiO_2 and kyanite, sillimanite, and andalusite are trimorphs of aluminum silicate (Al_2SiO_5). Paramorphs have also been called transformation pseudomorphs (Strunz, 1982), inversion pseudomorphs, and Umlagerungspseudomorphosen (Metz, 1964). Strunz (1982) rejects this latter usage.

a. Enantiotropic (reversible); e.g., alpha to beta quartz. If alpha quartz is heated over 573°C it inverts to beta quartz; when cooled below 573°C it reverts back to alpha quartz.

b. Monotropic (irreversible); e.g., graphite after diamond or low (alpha) quartz after stishovite, coesite, or tridymite.

    The first (a) involves only minor changes in the crystal lattice. The second (b) is characterized by substantial reorganization of the constituent atoms. Therefore, the first is reversible and the second is essentially irreversible.

2. Metamict, e.g., low zircon after high zircon. Metamict crystals are not traditionally classified among pseudomorphs, but some authors do include them including Strunz (1982). A metamict mineral is one that, although originally crystalline, has had its crystal structure damaged by the type of radiation known as alpha particles. Alpha particles consist of two protons and two neutrons making them identical to the nucleus of a helium atom. They do not have much penetrating power, but they can do a lot of damage in the short distance they travel. They can turn a crystal that was originally crystalline with the properties characteristic of its crystal structure and composition into an opaque amorphous mass with much lower properties, such as refractive index and specific gravity, than those possessed by the original crystal. Even a small amount of structural damage by alpha particles can change the physical and optical properties of the host crystal. There are at least 50 minerals known to occur in a metamict state.

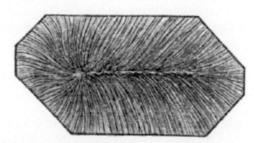

FIGURE 1.5 Needle-like hornblende crystals pseudomorph after augite single crystal (Scheerer, 1853).

3. Exsolution pseudomorphs. Also called Entmischung-Pseudomorphosen. These were not known to the early giants of the study of pseudomorphs such as Blum (1843, 1847, 1852, 1863, 1879), Landgrebe (1841), and Scheerer (1853, 1854a,b,c,d, 1856, 1953) (Fig. 1.5). The first exsolution pseudomorph was described by Pelikan (1902). He discovered that some "ilmenite" crystals were actually rutile (titanium oxide) and magnetite (iron oxide) that had separated out of homogeneous ilmenite (iron titanium oxide), which had formed at higher temperatures. The crystals became unstable and separated into laths of rutile and magnetite as the crystal slowly cooled. The study of exsolution pseudomorphs has become very important in recent years in the study of ore deposits.

4. Alteration pseudomorphs (processes involving chemical reactions). Note that these are very similar to those used by (Blum, 1843, 1847, 1852, 1863, 1879; Landgrebe, 1841) and other European authors a century and a half ago.

a. Loss of a constituent; e.g., copper after cuprite or azurite

b. Gain of a constituent; e.g., malachite after cuprite, gypsum after anhydrite

c. Partial exchange of constituents; e.g., goethite after pyrite, galena after pyromorphite

d. Total replacement; e.g., quartz after calcite, barite, or fluorite

e. Patinas and other alterations of artifacts; e.g., the Statue of Liberty whose mineralogy has been carefully studied and found to be a complex mixture of mostly rare secondary copper minerals (Nassau et al., 1987).

5. Replacement pseudomorphs. The process involves complete or partial solution and chemical precipitation of a new substance.

a. Infiltration; e.g., silicified wood such as the huge petrified logs in the Petrified Forest in Arizona. The silicification of wood has traditionally been described as a process where silica replaces the substance of the wood. Recently, a young German scientist, Michael Landmesser, has convincingly demonstrated that the silica does not replace the woody substance but only infills all of the pore space. The original work was published in an Extra Lapis in 1994 that was devoted to petrified wood (Landmesser, 1994) and in Chemie der Erde (Landmesser, 1995, 1998).

b. Replacement; e.g., native copper after aragonite from Corocoro, Bolivia.

6. Encrustation pseudomorphs (perimorphs).

    "Pseudomorphs, epimorphs, perimorphs. Molecules of a crystal can be replaced by molecules of another material without altering the form of the crystal. Thus a crystal of pyrite may be changed to goethite. If the change is more or less complete, a pseudomorph of one mineral has been formed on the original mineral; if the change is superficial, as for example, when only the periphery of the pyrite crystal has been change to goethite, an epimorph of goethite has been formed on the pyrite. If a crystal of one mineral is encrusted by another, and the original mineral is later leached out, a perimorph is formed" (Cissarz and Jones, 1931). There is much confusion in the literature and even on museum labels between epimorphs and perimorphs.

   a. Epimorphs.

   b. Perimorphs sensu strictu: where an original mineral is encrusted by a second and then the first mineral is leached away leaving a hollow shell in the form of the original mineral. They have also been called encrustation pseudomorphs or Umhüllungspseudomorphosen.

   c. Narben: In German die Narbe means scar. Swiss mineralogists use the term for the scars or impressions left by a now vanished minerals on the faces of another crystal. The most famous examples are smoky quartz crystals from Switzerland that show the impressions of now vanished fluorite crystals.

   d. Molds and casts including polyhedroids. Polyhedroids (also called box quartz, rectanulos (Port.), lattice quartz, Polygonachat (Ger.), quartz interstices, Polyedrische Quarz Drusen (Ger.), Pseudoachat (Ger.), Phantomachat (Ger.), radiate bladed quartz, geometric geodes, box geodes, triangular agate, polygonal umgrentzte Achate (Ger.), pegmatite agates, quasicrystals, angle-plated quartz, Zwickelfülling (Ger.) (Zwickel is a German term for space between crystals), Catazeiras (Port.), Paraiba agate, Poly-Quarz (Ger.), and Poly-Hydrolite-Achat (Ger.) have also been erroneously described as pseudomorphs after feldspar or calcite. Polyhedroid is the preferred term. They actually are silica infillings between now vanished, thin, flat plates of calcite that intersected at random angles. They could be called pseudo-pseudomorphs.

7. Fossils. Haüy (1801) coined the term pseudomorph to apply specifically to mineral replacements of plants and animals. Just how and when the term came to be used in its present sense is something we have not been able to discover. Suffice it to say that by the middle of the century before the last century, Haüy's term had been changed to essentially the modern sense.

   a. Petrifactions.

   b. Molds (of animals or plants).

   c. Casts.

8. Complex combinations of the above.

   There is very little in the natural world that is actually totally simple and straight forward. This category is for those more interesting cases where more than one process is involved. Especially in ore deposits, the sequence of events can be quite complex, but unraveling that sequence can be quite valuable in understanding a particular deposit or prospecting for a new one.

9. Unknown, obscure, fake, and/or highly controversial.

   a. Unknown. Most pseudomorph labels should probably read "pseudomorph after undetermined mineral, possibly…"

   b. Obscure. Another degree of A

   c. Fake

   d. Highly controversial. Identifying the present composition of a pseudomorph is pretty straight forward. Identifying what the now vanished mineral(s) was (were) is deductive, which more often than not leads to more than one possibility.

MODERN DAY INSIGHTS

Deep in the Earth's crust and in the mantle, under high temperatures and/or pressures, transformations can take place via solid-state reactions. However, on the Earth's surface and at typical sediment temperatures, solid-state processes are kinetically slowed down and their influence to most mineral transformations can be considered insignificant. Under these conditions, the presence of aqueous solutions offers an efficient mass transfer medium. Thus mineral transformations can occur at a substantially accelerated through the combination of two processes: the dissolution of the reactant phase and the crystallization of the new one. These transformations frequently show the characteristics of mineral replacements (Putnis, 2002), i.e., the volume and the shape of the reactant crystals are preserved during the transformation. This implies that the dissolution of the reactant and the crystallization of the product phase (or phases) necessarily occurs simultaneously and at coupled rates (Harlov et al., 2007; Putnis et al., 2007b). Moreover, the formation of pseudomorphs normally occurs (Putnis et al., 2006; Putnis and Putnis, 2007; Putnis et al., 2007c; Sánchez-Pastor et al., 2007). Under certain circumstances, pseudomorphization can be extremely faithful to both the shape and the surface features of the original crystal, which are precisely preserved (Fernandez-Diaz et al., 2009).

O'Neil and Taylor (1967) nearly 50 years ago proposed that the cation and oxygen isotope exchange between alkali feldspars and aqueous chloride solutions occur through a process of fine scale dissolution and redeposition. Instead of solid-state diffusion, this process, termed interface-coupled dissolution—reprecipitation (Putnis and Putnis, 2007), was later suggested to be the primary mechanism of mineral replacement reactions in the presence of a fluid phase (Nishimura et al., 2004; Pollok et al., 2011; Putnis, 2002, 2009b; Putnis and John, 2010; Putnis and Mezger, 2004; Xia et al., 2009a). Recently, a significant number of experimental studies confirmed that reactions between minerals and a fluid phase, also comprising simple cation exchange reactions, frequently comprise a pseudomorphic replacement via an interface-coupled dissolution—reprecipitation process (e.g., Geisler et al., 2005a; Harlov et al., 2005; Hellmann et al., 2003; Labotka et al., 2004; Niedermeier et al., 2009; Putnis and Putnis, 2007; Putnis et al., 2005). The process of pseudomorphism is understood to begin with superficial congruent dissolution of the parent phase resulting in a thin fluid film which is supersaturated with respect to a more stable phase. Ensuing epitaxial precipitation of the product phase onto the surface of the parent phase preserves its shape and crystallographic orientation, whereas the reaction continues into the parent phase. The driving force for such a coupled process is a solubility difference between the parent and the product phase, with the product phase having the lowest solubility in the fluid. This and possibly a smaller molar volume of the product phase results in the formation of porosity within the product phase. If the connectivity of this porosity is high enough the replacement can progress as the fluid is able to reach the reaction interface. The main features of such replacement process are a sharp chemical interface between the parent and the product phase, the preservation of the external dimensions and crystallographic orientation of the parent phase, and the development of porosity within the product phase (Putnis, 2002, 2009a,b; Putnis and Putnis, 2007).

Interface-Coupled Dissolution—Reprecipitation

In order to gain a better understanding of the mechanisms involved in the pseudomorphic replacement of one mineral by another mineral numerous papers have been published reporting on experimental work related to this topic. Below are a number of the most important examples showing the mechanism of interface-coupled dissolution—reprecipitation involving the replacement of one nickel sulfide by another sulfide, calcium carbonate by another carbonate, or by a calcium phosphate like apatite, as well as a few others.

Violarite, $FeNi_2S_4$, can be found in abundance in the supergene alteration zones of numerous massive and disseminated Ni sulfide deposits, where it replaces primary nickel sulfide minerals such as pentlandite (Misra and Fleet, 1974; Nickel, 1973). The nickel deposits of Western Australia's Yilgarn Craton have deep weathering profiles and supergene violarite forms a substantial amount of the ore in some of these deposits. Therefore, violarite is possibly the most economically significant member of the thiospinel mineral group. Violarite can furthermore form as a primary phase through exsolution during the cooling of pentlandite $((Fe,Ni)_9S_8)$ (Grguric, 2002). Understanding the thermodynamics and kinetics of the formation of violarite in the weathering profile is imperative to gain insight in alteration patterns in and around nickel deposits and has important consequences for ore processing. Supergene violarite is normally very fine-grained and reasonably porous and it has a poor response in the floatation systems used to treat various massive sulfide ores. Conversely, a quantity of violarite in the nickel concentrate helps smelting as the burning of violarite is a highly exothermic reaction (Dunn and Howes, 1996). Therefore, the processing of nickel sulfide ores might gain from a better insight in the conditions and mechanisms affecting the transformation of pentlandite to violarite. Putnis (2002) indicated that the porous and cracked texture displayed by supergene violarite point to a dissolution—reprecipitation reaction instead of a topotactic transformation. Pring et al. (2005) reported on the replacement of pentlandite by violarite in an ore concentrate (90 wt% pentlandite and 9 wt% pyrite with minor amounts of hydrotalc minerals, serpentine minerals and magnetite) using a water bath and synthetic pentlandite in the hydrothermal cell experiments with preliminary results under which the transformation progressed and on the kinetics and mechanism of the reaction. The water bath experiments proved that the transformation of pentlandite to violarite occurs at $T = 80°C$ within the pH range 3—5. About 20(4)% of the pentlandite transformed to violarite is 35 days. The value of the pH within the range 3—5 did not appear to considerably affect the transformation rate. The transformation of pentlandite was increased through the addition of small amount of $H_2S(g)$ to the Ar stream and 1 mL of a 0.1 M solution of $Fe(CH_3COO)_2(OH)$ to the acetic acid solution. Under these settings, 50(4)% of the pentlandite was transformed to violarite in 35 days. At $T = 120°C$ and $P = 3.5$ bars, the pentlandite to violarite transformation took place

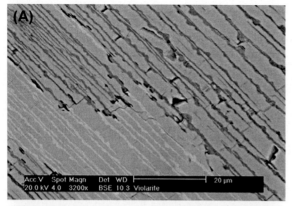

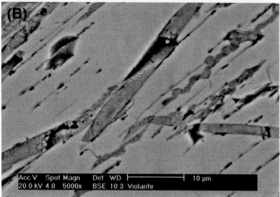

FIGURE 1.6 Back-scattered electron (BSE) image showing the partial transformation of the pentlandite/pyrrhotite to violarite/pyrrhotite. (A) The light mineral is unaltered pentlandite, the matrix is pyrrhotite, and the darker mineral is violarite. Note that where the pentlandite is transformed to violarite, the pyrrhotite matrix is cracked and pitted and that there is a gap of between 0.2 and 0.4 μm gap between the end of the pentlandite lamella and the violarite. (B) High magnification view of the newly formed violarite showing that the violarite is finely cracked and pitted giving it a porous texture (Pring et al., 2005).

over 3 days; after that time the pyrrhotite in the ore concentrate started to transform to marcasite or pyrite showing substantial dissolution of the pentlandite and pyrrhotite releasing $H_2S(aq)$ into solution in the course of the experiment. The cell parameters of pentlandite, violarite, and pyrrhotite did not change significantly over the course of the reactions, indicating no compositional readjustment during the transformation. BSE images illustrated that the pentlandite lamellae were increasingly transformed to violarite (Fig. 1.6). Where pentlandite is transformed to violarite, the pyrrhotite host was cracked and fractured allowing fluid flow through the matrix. It also showed a gap of between 200 and 400 nm between the end of the pentlandite lamellae and the violarite. Higher magnification images exposed the finely cracked and pitted texture of the secondary violarite. This texture is comparable to that observed by Grguric (2002) for supergene violarite. BSE diffraction showed the violarite to be very fine grained and not a single crystal. This study was the first to replicate experimentally, under mild hydrothermal P–T conditions, the reaction of supergene oxidation of pentlandite to violarite. The scanning electron microscopy (SEM) imaging proved unmistakably that the transformation mechanism is dissolution–reprecipitation (Putnis, 2002) and not a topotactical leaching of metal from pentlandite. The reaction front used only a small volume ($<<1\ \mu m^3$) of fluid. The pentlandite to violarite reaction is complex, contingent upon a great number of solution parameters, including pH, oxidation/reduction potential, and speciation and concentration of sulfur, iron, and nickel in solution. Available thermodynamic properties (Warner et al., 1996) suggested that, under acidic conditions, the reaction happens under very reducing conditions ($fH_2(g)$ c. 0.2 bars). A general reaction using a Fe sulfide (e.g., the thermodynamically stable pyrite) as an Fe sink, $H_2S(aq)$ as the aqueous sulfur species and oxygen as an oxidant can be expressed as:

$$\text{Pentlandite} + 2.75\ O_2(g) + 5.5\ H_2S(aq) = 2.25\ \text{violarite}$$
$$+ 2.25\ \text{pyrite} + 5.5H_2O \tag{1.1}$$

Following this equation, the transformation of pentlandite to violarite will not be pH dependant. Nevertheless, SEM work does not provide evidence for the formation of an iron sulfide coupled with the violarite precipitation. Therefore, it can be assumed that the excess iron was removed in the solution:

$$\text{Pentlandite} + 1.625\ O_2(g) + H_2S(aq) + 4.5\ H^+ = 2.25\ \text{violarite} + 2.25\ Fe^{2+} + 3.25\ H_2O \tag{1.2}$$

This reaction is driven by a more acidic pH and by high $H_2S(aq)$ concentrations. The equilibrium solubility of Fe and Ni at 80°C for conditions where pyrite, pentlandite, and violarite coexist in the presence of 0.2 molal of the Na-acetate–acetic acid buffer was calculated to fluctuate from 58 ppm Fe and 1.14 ppm Ni at pH 3.62–48 ppb Fe and less than a ppb Ni at pH 5.18, and therefore is sufficient to describe the preferential leaching of Fe. In the experiments, where Fe^{3+} was added, it can be assumed that the oxidant was Fe^{3+}:

$$\text{Pentlandite} + 6.5\ Fe^{3+} + H_2S(aq) = 2\ H^+ + 2.25\ \text{violarite} + 8.75\ Fe^{2+} \tag{1.3}$$

Eq. (1.3) is driven by higher Fe^{3+} and $H_2S(aq)$ concentrations, which explains why the reaction happened faster when $Fe^{3+}(CH_3COO)_2(OH)$ and $H_2S(g)$ were added to the solution (Pring et al., 2005).

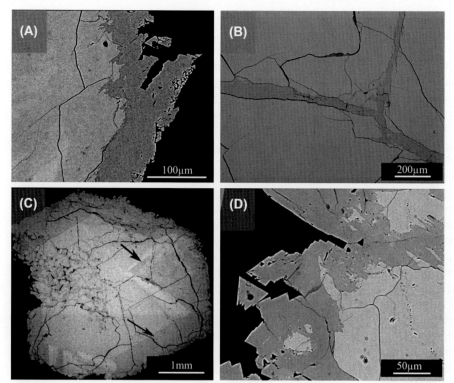

FIGURE 1.7 Back-scattered electron (BSE) images of cross-sections of samples treated in water and in solution S1 ($Na_2CO_3:CaCl_2 = 1$) for various periods of time. (A) Outer right part of a crystal reacted at 180°C showing replacement and overgrowths covering the external surface. (B) Central region of a crystal reacted at 200°C exhibiting some additional replaced regions along cracks. (C) Replacement observed along the outer parts of the crystal but also in the central left area. The *arrows* indicate the zones of heterogeneous Sr distribution, which are not replacement. (D) Close up of the left part of the crystal, showing the rhombohedral calcite shape of the overgrowths. In the right part of the image, pores can be observed leading away from the replaced regions (Perdikouri et al., 2011).

The experimental replacement of aragonite by calcite under hydrothermal conditions between 160 and 200°C using single aragonite crystals as a starting material was studied by Perdikouri et al. (2011). The original saturation state and the total $[Ca^{2+}]:[CO_3^{2-}]$ ratio of the experimental solutions were determined to have a decisive effect on the quantity and abundance of calcite overgrowths in addition to the degree of replacement observed within the crystals. The replacement process was accompanied by increasing formation of cracks and pores in the calcite, which resulted in extended fracturing of the original aragonite. The general shape and morphology of the starting aragonite crystal were preserved (Fig. 1.7). Whatever the origin of the aragonite, the thermodynamic prediction is that, under earth surface conditions, it has to transform to calcite with time, but the parameters that impede or promote the transformation are not well-known. Although the solid-state transition of aragonite to calcite is slow, even by geologic standards, dissolution and precipitation reactions may be fast and the transformation in an aqueous solution happens far too quickly to permit the preservation of aragonite in metamorphic rocks within the time scale of uplift.

It has been predicted that to preserve aragonite in high pressure rocks, it must enter the calcite stability field between 125 and 175°C and that its preservation necessitates the absence of a free aqueous fluid. The recognition of an aragonite precursor is usually based on textural indications as the aragonite morphology and microstructure is frequently conserved even after the replacement is completely finished. The impact of temperature and solution composition on the aragonite to calcite hydrothermal transition was examined for periods of 1 week–4 months at temperatures between 160 and 200°C with fluids of different compositions. The reaction formed two key types of calcite occurrences: (1) calcite overgrowths on the outside surfaces of the reacted crystal and on freshly uncovered surfaces from reaction-induced fracturing and (2) calcite sections created by replacement, forming rims, and/or spreading from fractures in the aragonite crystals. The first type can be distinguished texturally as euhedral or semieuhedral crystals either on the original crystal surface or in open fractures, whereas "replacement calcite" directly pseudomorphically replaces the aragonite at an irregular interface. Experiments run with pure water for 1 month exhibited no indication of any reaction at 160°C, whereas limited replacement of the aragonite by calcite was observed in run products from the tests at 180 and 200°C. Mineral replacement reactions have been suggested to take place by an interface-coupled dissolution–reprecipitation mechanism happening along a moving reaction interface (Putnis, 2009b). The absorption of ^{18}O-bearing carbonate in the calcite replacement reactions, observed with Raman spectroscopy, showed that the parent solid was dissolved and the new calcite phase was precipitated along the surface of the original crystal while taking up ions from the solution. This process is controlled by the composition of the fluid boundary layer formed by the dissolution of the original aragonite crystal and the concurrent crystallization of the calcite product at a moving front that initiates at the surface and partly spreads out through cracks in the aragonite crystal. These fractures were produced in the process of replacement and were not innate to the aragonite crystal. The hydrothermal replacement of aragonite by calcite did not yield perfect pseudomorphs, but can be labeled as "almost pseudomorphic" as used by Xia et al. (2009a). Throughout the reaction, the general outside dimensions of the original aragonite were preserved, except for the overgrowths. The preservation of the external morphology and dimensions describes the pseudomorphic replacement of one mineral by another (Putnis, 2009b). Two elements are essential for such a pseudomorphic replacement: (1) the dissolution of the parent and precipitation of the product must be coupled and (2) there have to be pathways for the solution to come in contact with the parent material (Putnis and Putnis, 2007). Both fracturing and porosity formation, ensuing from the mineral−fluid interaction, created the pathways for fluid to contact the aragonite crystal. The replacement of aragonite by calcite was strictly associated to the formation of pores and progressive fracturing, which separated the parent crystal into smaller domains. It is worth stating that, even though the aragonite−calcite phase transformation comprises an increase in molar volume, there is still porosity formation. The relative molar volumes and solubilities of the two phases are the two key factors that influence the porosity creation. Considering the difference in solubility between the two minerals [$Ksp_{(Aragonite)} > Ksp_{(Calcite)}$], more aragonite will be dissolved than calcite precipitated, thus creating the porosity and resulting in a loss of material to the fluid phase. This is anticipated for an interface-coupled dissolution−precipitation mechanism (Putnis and Putnis, 2007). Fracturing of the aragonite crystal may possibly result from the stress caused by the increase in molar volume of the solid during the replacement reaction (8.44%). Numerous fractures were situated around as well as within the replaced aragonite. The majority of the fractures in the crystal were close to or were opened from the aragonite−calcite interface in reaction to the replacement process. Fracturing is a common feature of pseudomorphic replacements comprising an increase in molar volume (Putnis, 2009b). These fractures offer pathways for mass transport within the crystal, allowing lateral dissemination of the reaction and consequently effect the replacement rate. This study suggests that the degree of replacement is governed by the composition of the starting solution (Perdikouri et al., 2011). As the replacement comprises the dissolution of aragonite, the starting solution possibly will effect the dissolution rate, which in turn will control the composition of the solution at the solid−fluid interface, where the calcite nucleates. Yet, the amount of replacement was also associated with the degree of associated calcite overgrowth precipitation on the external surface of the original aragonite. An important difference between the overgrowth and the pseudomorphic replacements is that the overgrowth resulted in close to perfect rhombohedral calcite crystals with no obvious porosity, whereas the replacement calcite is porous. The nonporous overgrowths restrict access of the fluid to the aragonite and limit further replacement within the aragonite crystal, i.e., they behave as a barrier layer. Therefore, if calcite overgrowths totally covered the external surface of the aragonite, the unreacted aragonite would become isolated from the solution preventing further reaction. The variation in the solution stoichiometry had a decisive part in the growth rate of calcite and the extent of replacement. When a solution with equimolar $Ca^{2+}: CO_3^{2-}$ was used, the progressing calcite crystals incompletely sealed off the parent material, likely reducing or preventing further transformation due to their size. By altering the solution composition, the size of both the overgrowths and the crystals growing within the aragonite was successfully changed. Preceding theoretical and experimental studies have proven that a deviation from stoichiometric calcium to carbonate ratios in the solution considerably reduced the nucleation and growth rate of calcite even at constant supersaturation. In this study, changing the ratios of calcium to total inorganic carbon of the experimental solutions reduced both the amount and size of the calcite overgrowths. This permitted the solution to access the aragonite−calcite interface and further enable the replacement process. When solutions with calcium to total inorganic carbon concentration ratios = 10^{-3} or 10^3 were used, the reaction advanced much faster than in the stoichiometric solutions and was even more effective when a Na_2CO_3 solution was used. In this situation, the only source of Ca^{2+}, which would produce a local supersaturation at the fluid−mineral interface, is from the dissolution of aragonite. The porosity in the replacement calcite indicates that, under these conditions, the aragonite−calcite replacement progresses by an interface-coupled dissolution−precipitation mechanism and results in a more perfect pseudomorph. A high amount of ^{18}O incorporation was found in the overgrowths ($\approx 93\%$). Some replaced regions in the core of the crystal showed a significantly reduced ^{18}O content (down to $<10\%$), much lower than the ^{18}O composition of the solution after reaction ($\approx 85\%$), demonstrating that the solution in these regions had restricted contact with the bulk reservoir during the reaction time. The composition of the solution changed locally with the advancement of the reaction, with restricted fluid advection or diffusion between the inner aragonite and the bulk. The calcite overgrowth may well have inhibited exchange between the nucleating (or replacement) calcite and bulk solutions at the time (Perdikouri et al., 2011). Analysis of the crystallographic orientation of the product calcite using electron backscatter diffraction (EBSD) exhibited little

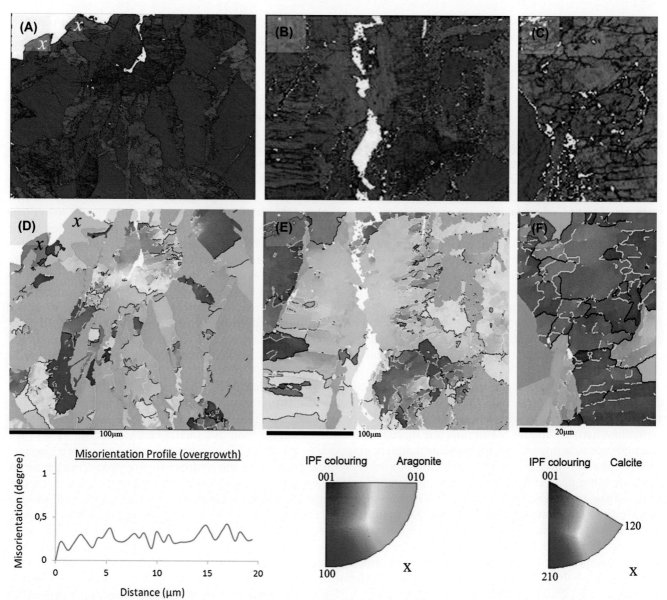

FIGURE 1.8 Electron backscatter diffraction (EBSD) maps of three representative areas of the sample treated in solution S1 for 1 month at 200°C. (A–C) Maps depict distribution of aragonite (blue) and calcite (red) (upper row); note the location of nonindexed points (white) along grain, phase, and subgrain boundaries as well as cracks. (D–F) EBSD maps of same areas as A–C showing crystallographic orientations based on the shown inverse pole figure coloring key (insets). Subgrain boundaries between 5 and 10 degrees misorientation are shown in yellow, grain boundaries with above 10 degree misorientation are shown in black. The parent aragonite is represented by the homogeneous orange color, whereas the calcite regions are made up of domains with very different crystallographic orientations, indicated by their different coloring. In (F) a twin is seen in aragonite (light and dark orange). (A + D) Region A exhibiting polycrystalline calcite within the aragonite with very different crystallographic orientations. Some areas have a high density of grain and subgrain boundaries; the typical lack of subgrain boundaries and systematic change of crystallographic orientation within calcite overgrowth crystals (marked as "X" in D) is seen as the orientation of the calcite overgrowth is constant (green and light blue). In addition, we show a misorientation profile across the "green" calcite overgrowth crystal in the inset. (B + E) Region B with a crack running through the crystal with an area of rather uniform calcite orientation in the center and with extensive subgrain boundaries observed in the outer parts of the fracture. (C + F) Region C shows calcite crystals (light blue color) extending across both twins (light and dark orange). The presence of the aragonite twin did not affect the orientation of the calcite that formed during the replacement reaction (Perdikouri et al., 2013 © www.schweizerbart.de).

to no link between the two phases under the studied conditions, with calcite crystallites showing dominantly different crystallographic orientations in comparison to the crystallographic orientations of the aragonite and of neighboring calcite domains (Perdikouri et al., 2013). This is in contrast to earlier studies (Folk and Assereto, 1976; Frisia et al., 2000, 2002; Kendall and Tucker, 1973) where the aragonite morphology and microstructure was frequently preserved even after complete replacement by calcite. Most of the calcite in this study (Perdikouri et al., 2013) showed wide-ranging grain and subgrain boundaries, and different orientations relative to both the original aragonite and to neighboring calcite grains. Minor relict regions of aragonite, located within the new calcite domains, have minor misorientations compared to the original aragonite, maybe due to rigid body rotation subsequent to the formation of multiple subparallel generations of fractures connected to the replacement mechanism. The overall lack of crystallographic relationships between aragonite and calcite in the samples indicates that in most parts of the crystal the dissolution and precipitation steps in the replacement were spatially uncoupled and that nucleation of calcite may well have occurred in open space, within the reaction-generated fractures, instead of epitaxially on aragonite surfaces. Later growth of calcite could then continue to replace dissolving aragonite with a sharp interface between parent and product. The large range of crystallographic orientations found in the calcite grains also points to independent nucleation events. The existence of subgrains within the calcite replacing aragonite could be related to stresses during growth but more work is necessary to determine the origin of the stresses and how transmission of these stresses into the host aragonite results in fracturing. The fractures created by the reaction are randomly spaced and their orientation is not crystallographically controlled. Fracturing is most likely connected to a combination of mechanisms (e.g., fracturing due to lattice expansion, fracturing due to fluid-assisted dissolution−precipitation, hydraulic fracturing due to local fluid pressure gradients), mostly involving stresses created by the volume increase and effects from the fluid mediated dissolution−precipitation process as well as local isolation of fluid pockets from the external fluid reservoir (Perdikouri et al., 2013) (Fig. 1.8).

The replacement of aragonite and calcite by apatite is a process that takes place naturally during diagenesis, chemical weathering, and hydrothermal reactions and is artificially endorsed in medical disciplines for the application of the product material as a bone implant. Aragonite and calcite single crystals can easily be converted into polycrystalline hydroxyapatite pseudomorphs through a hydrothermal reaction in a $(NH_4)_2HPO_4$ solution. SEM of the reaction products proved that the transformation of aragonite to apatite resulted in the formation of a sharp interface between the two phases and the formation of intracrystalline porosity in the hydroxyapatite phase. Furthermore, EBSD imaging proved that the c-axes of the apatite crystals were primarily oriented at right angles to the reaction front with no crystallographic relationship to the original aragonite lattice. Nevertheless, the Ca isotopic composition of the parent aragonite was preserved in the apatite phase. Hydrothermal experiments using phosphate solutions with water enriched in ^{18}O (97%) in addition showed that the ^{18}O from the solution was incorporated into the apatite structure. The textural and chemical results are pointing to a coupled mechanism of aragonite dissolution and apatite precipitation occurring at a moving reaction interface. The hydrothermally transformed crystal perfectly conserved the habit of the original aragonite, but its chemical composition and structure totally changed. The optical transparency of the initial aragonite vanished in the new phase. A comparable result was achieved with calcite crystals. The product crystals were strong and at all times exactly preserved the morphology of the original calcite crystal (Fig. 1.9). The results of X-ray diffraction (XRD) analysis proved that the new phase in both cases consisted of stoichiometric hydroxyapatite. SEM images of cross-sections of partially replaced aragonite samples exhibited polycrystalline hydroxyapatite and progressed from the initial aragonite surface through the creation and growth of elongated hexagonal crystals. This is more obvious in the back-scattered electron (BSE) images of polished cross-sections that also exhibit a sharp micrometer-size margin between the original and the product phase in addition to a high porosity in the product phase. The high porosity is also shown by the milky-white color of the crystals, a

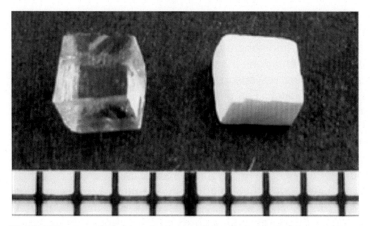

FIGURE 1.9 Single calcite crystals before (left) and after (right) the reaction. The morphology of the treated crystals is accurately preserved and the only macroscopically observable difference is the loss of transparency (Kasioptas et al., 2011).

consequence of the scattering of the light by the abundant pores in the hydroxyapatite product. Similar results were obtained with calcite as the starting crystal. The key difference concerning the two calcium carbonate polymorphs, at the studied temperature and composition of the solution, was that calcite at all times altered to apatite quicker than aragonite. The macroscopic and microscopic textures of the reacted samples are in line with those characteristically detected in fluid-induced pseudomorphic transformations (Putnis, 2009b). In a partially replaced crystal where the reaction was stopped in the initial stages (reacted for ≈3 days) only a few microns of apatite replacement was observed. In this case there was a noticeable gap at the boundary between the two phases, in addition it was obvious that the replacement front was not parallel to the initial surface of the apatite parent, instead exhibited a scalloped morphology. The fine-grained acicular apatite grains exhibited a preferred crystallographic orientation with the c-axes (matching the long axis of the needles) oriented nearly at right angles to the original crystal faces of the aragonite crystal. The dissolution of calcium carbonate in the phosphate solution liberates calcium ions into the solution, causing the development of supersaturation in regard to a less soluble calcium phosphate phase, apatite, in a thin fluid film at the reaction interface. As soon as a critical value of supersaturation has been reached in the interfacial fluid film, nucleation can occur at the start by crystallizing apatite on the surface of the aragonite. The reaction then proceeds by further dissolution and precipitation at the reactive interface, which proceeds through the sample by diffusion of phosphate ions in the direction of the interface through the pores formed in the apatite rim (Kasioptas et al., 2011, 2008) (Fig. 1.10). The replacement of a natural carbonate rock (Carrara marble) by apatite was used as a model to study the role of fluid chemistry in replacement reactions, concentrating on the mineralogy, chemical composition, and porosity of the apatite by Pedrosa et al. (2016). Carrara marble was reacted with di-ammonium phosphate solutions ((NH_4)$_2$$HPO_4$), together with and without different salt solutions (NH_4Cl, $NaCl$, NH_4F, and NaF) at various ionic strengths, at 200°C and autogenous pressure. The reaction in every sample

FIGURE 1.10 Scanning electron microscopy (SEM) images of two partially replaced aragonite samples. The dark core is aragonite, whereas the lighter rim is the new apatite phase in all cases. (A) Scanning electron image of a cleaved sample reacted at 200°C for 2 weeks. (B) SE image of close up on the left side of the sample showing the polycrystalline apatite product. (C) Back-scattered electron (BSE) image of polished cross-section of a sample reacted at 180°C for 2.5 weeks. (D) BSE image of close up on the left corner of the sample illustrating the porosity present in the apatite rim and the complex orientation of the apatite crystals (Kasioptas et al., 2011).

produced in pseudomorphic replacements and showed the features of an interface-coupled dissolution—precipitation mechanism. Increasing the ionic strength of the phosphate fluid amplified the replacement rates. With a constant concentration of phosphate, replacement rates were reduced with the addition of NH_4Cl and NaCl and increased considerably with the addition of NaF and NH_4F. The addition of different salts produced particular porosity structures caused by the formation of different phosphate phases. Chloride-containing fluids exhibited a higher level of fluid percolation along grain boundaries. After the reaction, the external dimensions and the cubic habit of all samples were preserved. BSE images and energy dispersive X-ray analysis (EDX) analyses of cross-sections of the reacted samples exposed that after 4 days some samples were in part replaced whereas others were completely replaced. The BSE images also showed that the internal microstructure of the marble was replicated in the new phase, including the shape and position of grain boundaries, and numerous fractures of the initial marble (Fig. 1.11). Significantly, the replacement apatite enclosed freshly formed porosity. In a closed system (such as in batch experiments) the replacement reaction produces local variations in the composition of the bulk fluid, but most significantly, in the newly formed pores of the sample itself. A changing solution composition as the reaction interface progresses through the reacting crystal, permitting diffusion of elements in the fluid phase, can also be liable for the heterogeneity of the end products. Analogous examples of compositional and phase variability caused by replacement reactions have been described for the replacement of calcite by both dolomite and magnesite (Etschmann et al., 2014) as well as in apatite replacement in the presence of As in solution (Borg et al., 2014). The variation in scale of the porosity in each case may indicate the influence of the fluid composition on the nucleation density of the apatite, with the presence of NaF producing a lower nucleation density, larger crystals and coarser porosity whereas the presence of NH_4F produced a high nucleation density, finer crystals and porosity on a considerably finer scale. Using the original Carrara marble (calcite 99.7%) density calculated as pure calcite (2.71 g/cm^3) and the density of fluorapatite (FAP) as 3.20 g/cm^3, the porosity of most of the samples reacted to FAP could be calculated. The relative molar volume difference between calcite and FAP is -14.7%. This was in line with the amount of porosity formed in the samples reacted with 0.5 M NaF in the phosphate fluid. Samples reacted with higher F-concentration in the fluid had higher porosity and higher but more variable F counts in the apatite. This indicates that samples reacted with fluids containing a higher F concentration are close to end member FAP, whereas samples reacted with lower F content are probably replaced by hydroxy-FAP ($Ca_{10}(PO_4)_6OH_xF_{2-x}$) that would be somewhat less porous because the relative volume difference between calcite and hydroxyapatite (-13.4%) is lower than between calcite and FAP (-14.7%). The replacement reaction progressed from the surface into the center of the sample with virtually no visible effects of the presence of grain boundaries. This was unlike the samples reacted with chloride salts with the reaction taking place along grain boundaries first. This indicates that variations in wetting properties of fluids of different composition can influence the preferred reaction path taken, but can also be associated with the much higher reaction rate in F-bearing fluids and the crystallization of FAP. In the course of the replacement of the calcite by calcium phosphates the carbonate may possibly be released to the fluid phase or reincorporated in the newly formed phases. The observation that the pH of the fluids at the end of the experiments did not increase, as would be anticipated if substantial amounts of carbonate ions were dissolved, suggests that carbonate released during the dissolution of calcite probably has been incorporated into the structure of the product phases (Pedrosa et al., 2016).

The mechanism and the kinetics of this replacement were also studied by Kasioptas et al. (2010) but in this case by using biogenic aragonite (cuttlebone of the *Sepia officinalis*) as a starting material and reacting it with di-ammonium hydrogen phosphate solution. Isothermal experiments were carried out at temperatures up to 190°C. SEM proved that the fine structure of the cuttlebone was flawlessly preserved even after aragonite had been totally changed to apatite (Fig. 1.12). The reaction product was a carbonated apatite, containing traces of strontium. Apart from carbonate, traces of Sr from the cuttlebone may have fit in the Ca sites in the apatite, although for Sr incorporation less than 1.5% this does not result in substantial changes in the apatite lattice parameters. Analogous replacement experiments of inorganic aragonite crystals that also produced a pseudomorph contained only apatite as a product phase. The apatite in that case was polycrystalline and formed elongated hexagonal needles, in contrast to the morphology exhibited by the cuttlebone apatite. Calcite corals were shown to change directly to tricalcium phosphate (TCP). The formation of various phases of calcium phosphate are determined by temperature, pH, solution composition and many other conditions. In the experiments with cuttlebone the formation of apatite was always complemented by the formation of TCP. The formation of hydroxyapatite can be complemented by a second calcium phosphate solid phase, depending on bulk composition. The apatite in these experiments cannot be described as well-defined crystals, and if some are found, range from nano- to micrometer in size as overgrowths whereas apatite that replaces inorganic aragonite under the same conditions forms recognizable, micro- to millimeter-sized

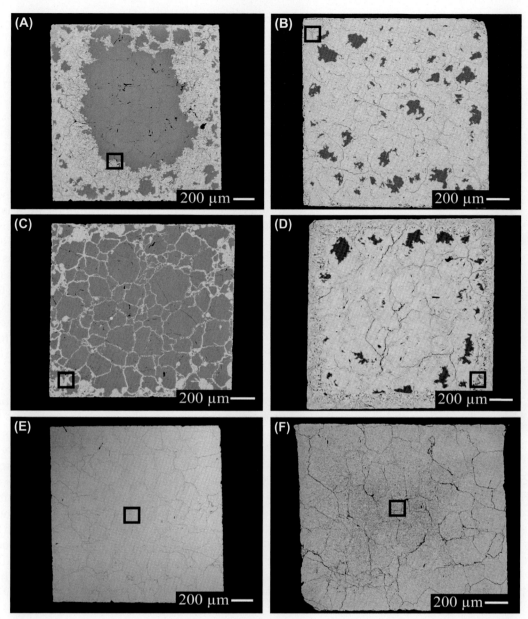

FIGURE 1.11   Back-scattered electron (BSE) images of cross-sections of Carrara marble cubes reacted for 4 days at 200°C, using different solutions: (A) 0.5 M $(NH_4)_2HPO_4$; (B) 2 M $(NH_4)_2HPO_4$; (C) 1 M $(NH_4)_2HPO_4 + 0.5$ M NaCl; (D) 1 M $(NH_4)_2HPO_4 + 0.01$ M NaF (similar result with 0.01 M NH_4F); (E) 1 M $(NH_4)_2HPO_4 + 0.5$ M NH_4F; (F) 1 M $(NH_4)_2HPO_4 + 0.5$ M NaF. *Darker gray areas* correspond to the unreacted marble and *brighter gray areas* to the replacement product (Pedrosa et al., 2016).

hexagonal crystals. The activation energy E_a implies a change in the mechanism with reaction advancement. A potential explanation is that in the initial stages of the reaction the overall rate is controlled by the large surface area of the cuttlebone. As the reaction progresses the reaction interface moves through the aragonite, necessitating larger scale mass transport of phosphate and carbonate through the growing thickness of the apatite product. This indicates that the experimental activation energy may be controlled by the surface reaction in the initial phases and increasingly on mass transfer as the reaction progresses. When the cuttlebone is placed in a $(NH_4)_2HPO_4$ solution, the $CaCO_3$ reacts with the phosphate and calcium phosphate precipitates by replacing the initial mineral. This pseudomorphic process permits the initial aragonite chamber-like architecture to be preserved and displayed by the new apatite phase (Fig. 1.12). Comparable aragonite-hydroxyapatite conversions [Porites corals and Gastropod

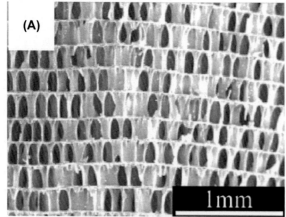

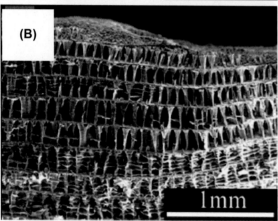

FIGURE 1.12 Secondary electron scanning electron microscopy (SEM) images of the initial cuttlebone showing the overall structure (A), the aragonite growth layers across the pillars. A reacted cuttlebone sample is shown in (B). The unique morphology of the cuttlebone is perfectly preserved even when it has been completely replaced by hydroxyapatite and a small fraction of tricalcium phosphate (Kasioptas et al., 2010).

(Abalone) nacre] have been suggested to involve a dissolution and recrystallization mechanism. It has furthermore been proposed that the replacement of cuttlebone by apatite could happen by diffusion controlled growth of needle-like crystals with main growth in the c-axis direction (Kasioptas et al., 2010).

The calcium carbonate replacement process is a pseudomorphic reaction as previous studies have also demonstrated (Eysel and Roy, 1975; Kasioptas et al., 2008; Putnis, 2009b). The reaction proceeds by successive dissolution and precipitation at a reactive interface, which migrates through the sample by diffusion of phosphate ions to the interface through the pores present in the apatite rim. It is significant to note that the detected gap at the interface between the aragonite parent and the hydroxyapatite product appears to be a general feature of replacement reactions where the crystallographic structure of the initial phase is different to that of the product phase. For example, a gap at the reaction interface has similarly been detected in the experimental replacement of self-irradiation-damaged pyrochlore by anatase and rutile (Pöml et al., 2007), in addition to the experimental replacement of magnetite by Fe disulfide (Qian et al., 2010). The cause for the formation of such a gap at the interface might be that after the original nucleation of the product phase onto the parent surface, further nucleation and/or growth of the product phase only happens on the by this time existing product grains because this form of nucleation-growth would be energetically more favorable. In this case this suggests that the Ca released through dissolution of aragonite must be transported through the interfacial fluid from the site of dissolution to the formed apatite crystals. Because of the 6% smaller molar volume of apatite compared to aragonite and because the overall volume of the aragonite sample is retained in the replacement reaction, the thickness of the gap ought to increase with increasing amount of replacement, or a larger porosity in the hydroxyapatite should be detected, as is in fact seen in the core of the completely replaced crystal. The well-preserved morphology of the original aragonite also indicates that the rate-limiting step in the replacement is the dissolution of aragonite, and not the precipitation of apatite (Xia et al., 2009b). The rate of replacement of calcite is faster than that of aragonite, notwithstanding that calcite is probably the less soluble phase. Nevertheless, the molar volume of apatite is $\approx 6\%$ smaller than that of aragonite and $\approx 12.7\%$ smaller than that of calcite. The replacement of the carbonates by apatite is pseudomorphic in both cases, which suggests that the porosity in the apatite rim will be higher after the replacement of calcite. This could clarify the faster reaction rate, boosted by an increased rate of mass transport to and from the reaction interface through the apatite rim. Still, in these experiments the crystallographic orientation of the apatite is not determined by the crystallographic orientation of the aragonite. The direction of the long axes of the apatite needles (i.e., [001] of the apatite structure) is roughly at right angles to the initial aragonite surfaces, i.e., perpendicular to the reaction front. Apatite needles showing a particular crystallographic orientation with regard to the aragonite were merely a consequence of the orientation of the reaction front in that particular region. The crystallographic directionality of the apatite needles was furthermore not influenced by the occurrence of an aragonite twin, in the same way as the observations of Zhao et al. (2009) where the twinning of the initial calaverite did not affect the orientation of the freshly forming—Au fibers (Kasioptas et al., 2011).

Atomic force microscopy (AFM) research by Klasa et al. (2013) revealed that the dissolution rate of calcite (CaCO$_3$) was considerably reduced in the presence of (NH$_4$)$_2$HPO$_4$ in solution but not in the presence of Na$_2$HPO$_4$ (both solutions pH ~8) in comparison to that detected in pure deionized water, indicating that the NH_4^+ group could impede calcite dissolution. The AFM experiments were performed under rather different conditions of fluid flow over a free cleaved calcite surface and therefore the chemical composition of the fluid was kept constant, not permitting for fluid reequilibration. In the Carrara marble tests it is conceivable that the higher NaCl concentrations accompanied by PO_4^{3-} in solution effectively enlarged the threshold for supersaturation of the precipitating phase within the boundary layer. Throughout a replacement process, natural porosity of rocks, such as grain boundaries and fractures, plays a significant role in permitting fluid access within the rock (Pedrosa et al., 2016). This improves the accessible internal surfaces where replacement reactions are started. Fluid movement through freshly formed pores plays a significant role in the progress of a reaction front. If fluid movement was restricted, the reaction front may be halted. In that case, the replacement process would end. Fluid movement might also be restricted by the loss of permeability. Permeability is subject to the interconnection of pores formed in the product phase. The significance of interfacial fluid composition on replacement reactions is comprehensively described in Putnis (2009b). The results suggest that there are two types of fluid pathways: along existing grain boundaries and fractures and through newly formed connected pores. It has been hypothesized that when replacement rates are extremely low, the lack of freshly formed pores results in the motion of fluid along the grain boundaries, causing the replacement of interior grains (both surface and interior grains showed very thin replacement rims), as observed in the samples reacted with high chloride content solutions. In contrast, when replacement rates are high, the fluid travels through the newly formed pores as observed in the samples reacted in the presence of fluoride. The movement of the fluid preferentially through newly formed pores instead of fractures and grain boundaries could be related to wettability. The product and parent phases have different critical surface tensions that are the result of the chemistry of each solid phase, and thus have different surface tension or wetting properties (Pedrosa et al., 2016).

Apatite $(Ca_5(PO_4)_3(OH,F,Cl))$ is probably the most significant hosts of halogens in magmatic and metamorphic rocks and plays a special role during fluid—rock interaction as it integrates halogens (i.e., F, Cl, Br, I) and OH from hydrothermal fluids to form a ternary solid solution of the end members F-apatite, Cl-apatite, and OH-apatite. Kusebauch et al. (2015) presented an experimental study on the interaction of Cl-apatite with different aqueous solutions (KOH, NaCl, NaF of different concentrations, also doped with NaBr, NaI) at crustal conditions (400—700°C and 0.2 GPa), resulting in the formation of new apatite. Due to a coupled dissolution—reprecipitation mechanism new apatite crystallized every time as a pseudomorphic replacement of Cl-apatite. In addition, some experiments resulted also in the formation of new apatite as an epitaxial overgrowth. The composition of the new apatite was largely controlled by the complex features of the fluid phase from which it precipitated and was determined by the composition of the fluid, temperature, and fluid to mineral ratio. In addition, the replaced apatite exhibited a compositional zonation attributed to a compositional progression of the coexisting fluid in local equilibrium with the newly formed apatite. Replacement characteristics were divided into two separate groups on the basis of the appearance of the replaced apatite and the coexisting pores therein. Group A was described by a high concentration of nanometer to micrometer-sized rounded pores, which were occasionally elongated along the crystallographic c axis. The interface between unreacted Cl-apatite and replaced apatite was typically sharp, but could be diffuse in experiments at high temperatures (700°C) with NaCl and NaF solutions. The width of the replacement rim around unreacted Cl-apatite was even and was not related to the crystallographic orientation. Group A type replacement was found in every experiment with NaF solution and also at high temperature runs (>600°C) of NaCl and KOH solutions. Group B had bigger irregular pores (>5 μm), occasionally with a discrete shape demonstrating the prismatic habit of apatite crystals (negative crystal). The interface between unaltered Cl-apatite and replaced apatite was irregular in outline, exhibiting fingering and was generally obvious by the larger irregular pores. Porosity in the replaced apatite appeared to be limited to the large pores as no porosity in the submicrometer range was found with the techniques available. The outermost rim of replaced grains generally had a higher concentration of pores relative to the inner part of the replaced apatite near the interface (Fig. 1.13). Most KOH experiments and low temperature (<600°C) NaCl experiments exhibited this type of replacement. Irrespective of group A or group B replacement, porosity-free apatite crystallized either as an epitaxial overgrowth on replaced apatite or as euhedral crystals from the fluid in several reduction of the newly formed minerals, which increases the mobility of dissolved species (Putnis and John, 2010). The pseudomorphic replacement of Cl-apatite by OH-apatite or F-apatite produced porosity within the product phase because of variations in molar volume between the different apatite phases (Yanagisawa et al., 1999). Putnis (2002, 2009a,b) suggested an extra mechanism, specifically solubility differences between the precursor and replacing phases, to create porosity. Solubility differences appear to play a role in the course of apatite replacement as the detected porosity appears to be considerably greater than the 2.9% because of volume shrinkage. Indeed, results from the replacement experiments proved this additional mechanism to be

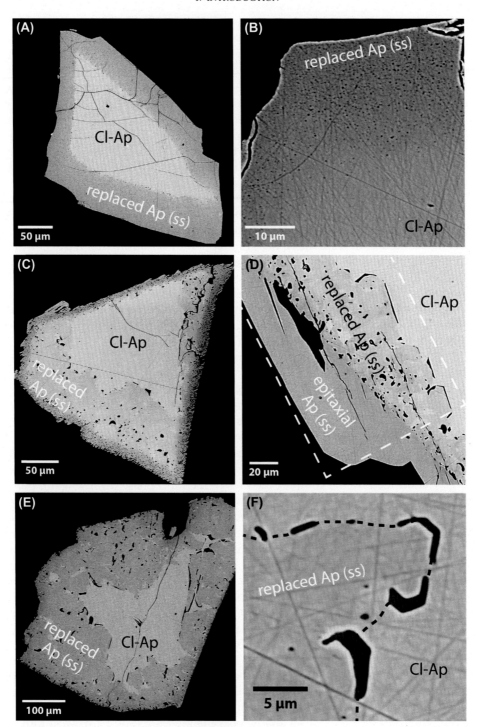

FIGURE 1.13 Back-scatter electron (BSE) images of replaced and epitaxial apatite after experiments; (A) type A replacement with NaF solution at 600°C showing lm sized pores and sharp interface between pristine and replaced apatite. (B) Type A replacement with NaF solution at 700°C with a diffuse interface. (C) Type B replacement with NaCl solution at 500°C showing segments of replaced apatite and large pores. (D) Detailed image of epitaxial and replaced apatite as in (C) showing segments of replaced apatite surrounded by pores tracing the prismatic shape of apatite. (E) Type B replacement with KOH solution at 500°C. (F) Detailed image of replaced apatite as in (E) with pores tracing the shape of apatite (Kusebauch et al., 2015).

of significance on a larger scale. It appeared that all elements of apatite (i.e., Ca, PO_4 groups) were dissolved at the replacement front and moved to the outside of the former crystal, where they precipitated again as epitaxial apatite. The former habit of Cl-apatite was still observed and marked by intersection of porous free epitaxial apatite to highly porous replaced apatite. The existence of epitaxial apatite furthermore points to an effective interconnectivity of the replacement front with the bulk fluid. During reaction of Cl-apatite with NaCl bearing fluid (10 wt%) at 600°C, two different replacement sectors at different reaction interfaces were found, i.e., an outer zone with small pores (type A replacement) and an inner zone with larger pores or cavities (type B). The shape of these cavities was adjoining partitions of replaced apatite in a perfect crystal shape and appeared to form a pore structure that was stimulating the creation of homogeneous apatite in these compartments. A coupling between morphology and composition of replaced minerals was also seen for fluid inclusions developed by a coupled dissolution—reprecipitation process in a closed system (Lambrecht and Diamond, 2014). The chemical composition of the replaced apatite in these compartments was strongly governed by the fluid composition in the cavities, suggesting that an internal zonation was initiated by a changing fluid because of continuing replacement in the direction of higher Cl- and/or lower OH-activities. A chemical zoning was similarly existing in replaced apatite of only type A replacement, usually in experiments with a low concentration NaF solution and with a NaCl solution at temperatures of 700°C. In these experiments, zoning followed an evolution in composition from center to rim and was not related to a sporadic distribution of sections of different composition. Chemical zoning in replaced apatite can have two potential causes: (1) diffusion controlled interchange of OH/F and Cl in the fluid between cavity and bulk fluid or solid-state diffusion in the replaced apatite or (2) depletion of OH and F from the fluid and concurrent enrichment of Cl either in the cavity/interfacial fluid only or in the bulk fluid system. In the apatite system, the width of replacement rims exhibited no reliance on time and was rather controlled by the amount of fluid in the system. In addition, the calculated rate of replacement and the development of the zoning were in the range of $10^{-15}-10^{-16}$ m^2/s, which is more than a few orders of magnitude slower than halogen diffusion in a free fluid at high temperatures. Consequently, it is impossible that the zoning within replaced apatite is formed by diffusion of Cl and OH in the fluid. Alternatively, solid-state diffusion in apatite is one to two orders of magnitude too slow at temperatures below 700°C ($<10^{-17}$ m^2/s) even when the crystal orientation (fastest parallel to c) is taken into account. Modeling diffusion profiles using extrapolated rates did not replicate the determined compositional profiles in replaced apatite and, consequently, diffusion is not the principal process to create the observed zonation. The second potential explanation of chemical zoning in replaced apatite is the development of apatite in local equilibrium with an evolving fluid composition. Because the replaced apatite has a lower solubility than the precursor apatite and is metastable in a slightly changed fluid, the chemical zonation should reflect the changing fluid composition during the experimental run. Thus only the first formed replacement apatite (outermost rim) will be in equilibrium with a fluid of starting composition. Despite a reliance on the amount of OH^- in the solution as showed by the zonation, the composition of replaced and epitaxial apatite was determined by primarily the experimental temperature. The partitioning behavior of halogens between hydrothermal fluid and apatite has important consequences for natural systems comprising apatite and for the use and understanding of partition coefficients. Even in a simple fluid—apatite system, quantitative relationships between fluid chemistry and apatite composition are intricate and not well understood as they are not merely a function of concentrations of dissolved species but of temperature, fluid composition, pH of the fluid, and fluid to mineral ratio as well. Qualitatively, this indicates that varying apatite compositions within one sample suite characterizes evolution of the fluid and the apatite chemistry reacts rather sensitively to these changes in the fluid composition. Particularly, in metasomatic settings, where fluid is pervasively penetrating the host rock and replacement is small, the fluid—rock interaction is similar to batch experiments at low fluid/mass ratios (rock dominated) and evolution of the fluid because continuing reaction with the host rock will be exemplified by a zonation of replaced apatite. In contrast, a setting where fluid is highly channelized will result in apatite with a more homogeneous composition as replacement happens under fluid dominated conditions with a fluid of constant composition (Kusebauch et al., 2015).

To reveal at what pressure and temperature and for what fluid compositions monazite may be induced to crystallize from FAP, the light rare earth elements (LREE)-enriched Durango FAP was metasomatized experimentally at temperatures between 300 and 900°C and pressures of 500 and 1000 MPa. Fluids used comprised pure H_2O, various NaCl, KCl, and $CaCl_2$ brines (salt/H_2O = 50/50, 30/70, or 10/90) and either 90/10 CO_2/H_2O or 40/60 CO_2/H_2O mix (Harlov and Förster, 2003), and 1 and 2 N HCl and H_2SO_4 (Harlov et al., 2005).

Monazite crystallized in the FAP + H_2O, FAP + 40/60 CO_2/H_2O, and the FAP + KCl brine experiments. At 900°C and 1000 MPa, monazite formed as inclusions within the FAP together with crystals externally on its surface (Fig. 1.14). Below 900°C, monazite formed only on the outside of the FAP, either as euhedral to semieuhedral crystals or as incomplete layers over smaller FAP grains. Monazite, in particular, at 900°C and 1000 MPa, was compositionally heterogeneous, in particular in regard to the Th content (ThO_2 = 4—38 wt%). Although the reactant FAP in the

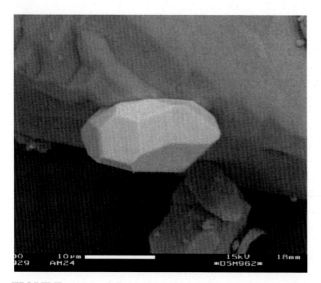

FIGURE 1.14 Monazite grain (monoclinic symmetry) nucleated on the surface of the fluorapatite (FAP) in H_2O-FAP experiments at 800°C and 1000 MPa (Harlov and Förster, 2003).

pure H_2O experiments stayed unzoned at lower temperatures, three coupled zones with varying (LREE + Si + Na) concentrations formed at 900°C. These regions crudely follow the border of the FAP surrounding a fourth zone or the center, approaching the original composition. Monazite inclusions were created only in the one zone where the LREE were depleted. In the NaCl brine experiments, Na replaced Si lost to the solution, which alleviated the LREE, and prohibited development of monazite. In the same way, the high activity of Ca in the $CaCl_2$ brine resulted in Ca to replace (LREE + Na) on the Ca site and depressed the formation of monazite. The FAP recrystallized to a fluor-chlorapatite that displayed oscillatory zoning, in particular regarding the LREE. The results from this study suggest that the existence of monazite inclusions and rim grains associated with FAP (1) can be metasomatically induced; (2) can provide a better understanding of the chemistry of the metasomatizing fluids; (3) can offer some evidence on the grade of the metasomatic overprint; and (4) may possibly point to the existence of one or more metasomatic events. In the fluor- and chlorapatite structure, the LREE mainly substitute on the Ca(1) site (ninefold coordination), whereas the (Y + heavy rare earth elements, HREE) favor the Ca(2) site (sevenfold coordination) (Fleet et al., 2000; Fleet and Pan, 1995, 1997; Hughes et al., 1991). Even though a number of coupled cation substitution reactions comprising (Y + rare-earth elements [REE]) are possible (Harlov et al., 2003; Pan and Fleet, 2002), in common FAP (Y + REE) are mainly charge-balanced through two coupled cation substitutions

$$Na^+ + (Y + REE)^{3+} = 2\,Ca^{2+} \tag{1.4}$$

and

$$Si^{4+} + (Y + REE)^{3+} = P^{5+} + Ca^{2+} \tag{1.5}$$

(Burt, 1989; Harlov and Förster, 2002; Pan and Fleet, 2002): These coupled substitutions are mutually charge-balance dependent and site dependent, in a way that the contributing cations have to be distributed 1:1 on the two specific sites accessible, i.e., P, Si, and S on tetrahedral sites, and Ca, Y, REE, and Na on the two Ca sites. During the metasomatism of apatite, the successive removal of Na and/or Si, but not the (Y + REE creates a charge imbalance and prompts the creation of monazite and/or xenotime through the following general mass transfer reactions:

$$[Ca_{5-2x},Na_x,(Y + REE)_x]P_3O_{12}F + x[2Ca^{2+} + P^{5+}]_{(fluid)} = Ca_5P_3O_{12}F + x(Y + REE)PO_4 + x[Na^+]_{(fluid)} \tag{1.6}$$

and

$$[Ca_{5-y}(Y + REE)_y][P_{3-y}, Si_y]O_{12}F + y[Ca^{2+} + 2P^{5+}]_{(fluid)} = Ca_5P_3O_{12}F + y(Y + REE)PO_4 + y[Si^{4+}]_{(fluid)} \tag{1.7}$$

(Harlov and Förster, 2002; Pan et al., 1993). This is supported experimentally, partial alteration of natural chlorapatite [(Y + REE)_2O_3 = 1–1.5 wt%] to hydroxyl-FAP in H_2O at 300–900°C and 500–1000 MPa formed monazite and xenotime as inclusions and rim grains in the metasomatized regions (Harlov et al., 2002b). In experiments involving H_2O-induced metasomatizatism of chlorapatite, monazite (besides xenotime) crystallized both as inclusions and border grains in sections of the chlorapatite earlier metasomatized to hydroxyl-FAP by H_2O-F fluids at temperatures and pressures as low as 300°C and 500 MPa (Harlov et al., 2002b) (Fig. 1.15). This observation indicates that the crystallization of monazite inclusions might as well depend on the composition of the apatite as on the composition of the fluid. Nucleation and crystallization of monazite on the surface of FAP is an example of autocatalysis as it increases dissolution of the surrounding apatite, which in turn increases formation of the monazite (Milke and Metz, 2002). Still, the monazite inclusions only formed in those parts of the FAP that also exhibited proof for having experienced what seems to be widespread, metasomatically induced obvious recrystallization. This result is in line with what was found in previous chlorapatite-fluid experiments where monazite and xenotime inclusions formed solely in those zones recrystallized to hydroxyl-FAP (Harlov et al., 2002b). These metasomatized areas, both in apatite

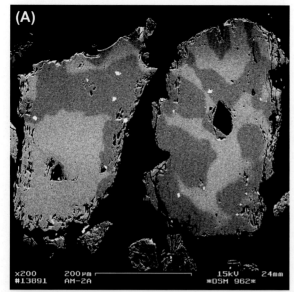

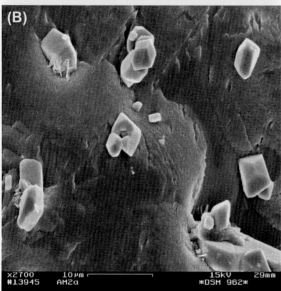

FIGURE 1.15 (A) Back-scattered electron (BSE) image of experimentally metasomatized (H_2O) chlorapatite from experiments at 900°C, 1000 MPa. (A) Shows both metasomatized regions (dark) with a scattering of a few relatively large monazite and xenotime inclusions as well as nonmetasomatized regions (light). *White spots* denote the presence of monazite and/or xenotime grains. (B) Grains of monazite (monoclinic symmetry) and xenotime (tetragonal symmetry), which have nucleated on the surface of a metasomatized area in chlorapatite in experiment at 900°C; 1000 MPa (Harlov et al., 2002b).

in addition to other minerals, have been inferred as replacement reactions that progress by means of a moving interface related to what is basically a simultaneous dissolution—reprecipitation process (Putnis, 2002). One of the results of such a progression is the formation of a large number, fairly homogeneously dispersed micropores in the metasomatized parts of the apatite. This result has been confirmed in other apatite-metasomatism studies concerning a range of apatites including an (LREE + Si + S)-enriched FAP (Harlov et al., 2003); chlorapatite (Harlov et al., 2002b; Yanagisawa et al., 1999); and fluor-, chlor-, and hydroxylapatite (Rendon-Angeles et al., 2000a,b,c). A second, and significantly more minor, influence on the microporosity is the minor volume decrease observed in those metasomatized areas of the FAP presenting a depletion in (LREE + Si + Na that could be ascribed to a britholite [$(Ca,REE,Na)_{10}(P,Si)_6O_{24}F_2$] component. Reduction in (LREE + Si + Na) therefore may be seen as a decrease in the britholite component causing a minor drop in volume because end-member britholite has a larger volume ($\approx 3\%$) than end-member FAP (Oberti et al., 2001). The impact of volume reduction to the microporosity was observed more intensely in the chlorapatite-fluid experiments of Harlov et al. (2002b). In that work, metasomatic change of chlorapatite to hydroxyl-FAP caused a negative volume adjustment of about 3%, which, when detected in situ, created a dramatic opening of pore space alongside the reaction front between chlorapatite and hydroxyl-FAP, in which and along which, both monazite and xenotime formed (Harlov et al., 2002b). The existence of micropores might aid both as local sites for monazite nucleation and crystallization along with providing a network that might let fluid move all through the metasomatized areas (Putnis, 2002). Fluid-aided transference of cations and anions through such a network in the FAP would significantly aid monazite crystallization by letting P and (Y + REE) to be transported into the developing monazite grain whereas Ca, Na, and Si would be moved out into the surrounding FAP and in the end out into the surrounding fluid. In such a situation, the monazite inclusions would at that point develop further by dissolution—reprecipitation detached from the surrounding FAP by a thin film of fluid. This explanation would clarify why monazite inclusions seem to form only in metasomatized regions that have experienced dissolution—reprecipitation and consequently a significant increase in their microporosity. Furthermore, such areas can also exhibit changes in their chemical composition, for instance an overall reduction in (LREE + Si + Na) as observed in the 900°C, 1000 MPa, FAP-H_2O experiment, or no clear reduction in the LREE but a reduction in (Na + Si) as exhibited in the 900°C, 1000 MPa FAP-KCl brine experiment. Prior studies of monazite and/or xenotime inclusions in both FAP (Åmli, 1975; Harlov et al., 2002a; Harlov and Förster, 2002; Pan et al., 1993) and chlorapatite (Boudreau and McCallum, 1990; Harlov et al., 2002b) have determined that these inclusions crystallized from the (Y + REE) budget accessible in reaction to the fluid-induced metasomatic change of the apatite. The results of this study (Harlov and Förster, 2003) endorse and build upon these conclusions. But, this obviously postulates that the FAP has a (Y + REE) content large enough (>0.2–0.3 total oxide wt%) that abundant monazite and xenotime crystals will form to turn out to be statistically significant, i.e., highly likely that they will be observed in apatite in a two-dimensional thin section under an optical microscope. In this respect, the fairly low bulk concentration of (Y + HREE) in the Durango FAP clarifies why xenotime was not created in any of the experiments. Analogous to what has been established for chlorapatite in Part I of this study (Harlov et al., 2002b), these experiments indicated further that monazite can crystallize over a rather large range of temperatures and pressures (300–900°C and 500–1000 MPa), but only in the company of a fluid with a restricted range of compositions. These results suggest that fluid chemistry, and not pressure and temperature, is the primary parameter enabling (or suppressing) the creation of monazite during FAP-fluid interaction. At this point H_2O has been determined to be the primary agent, whether as a pure fluid or otherwise diluted in some other constituent such as CO_2 or KCl. Therefore, whether or not monazite is accompanying with LREE-bearing FAP aids to put limits on the chemistry of the infiltrating fluids responsible for both the metasomatism of the FAP and the rock as a whole. One limitation, though, is that comparison between experiment and nature is a comparison between a short term (days to weeks) FAP-fluid experiment and FAP in nature exposed to comparatively more complex, high-grade fluids over time periods of 100,000s to millions of years under comparable P-T environments. Such a comparison suggests that though the results of these experiments imitate what is observed in nature, with the following implications, they cannot be used as a complete confirmation of the exact P-T fluid conditions under which these inclusions developed nor of their stability and/or probable progression over long periods of time. One chief difference with the chlorapatite study in Part I, though, is that monazite inclusions in FAP, in any case when pure H_2O, H_2O/CO_2, or a KCl brine is the metasomatizing fluid, can start to nucleate only at 900°C, whereas below this temperature, they form either as border grains or as layers over small FAP grains. This result is different from that observed in the chlorapatite-fluid experiments. In that case monazite (and xenotime) inclusions developed over the full 300–900°C and 500–1000 MPa temperature and pressure range. Whether this difference is only a function of either the apatite composition or the fluid composition (or both) is unclear at this stage. Nevertheless, in either case, formation of monazite and/or xenotime inclusions in either fluor- or chlorapatite seems to happen only in areas in the apatite that have been metasomatized through dissolution–reprecipitation and therefore represent regions with a high microporosity (cf. Putnis, 2002).

An experimental study of the interaction between gypsum (010) surfaces and aqueous solutions of Na_2CO_3 with different concentrations was performed by Fernandez-Diaz et al. (2009). This interaction led to the carbonation (i.e., the transformation into carbonate minerals) of gypsum crystals, which under ambient conditions exhibited the characteristics of mineral replacement and led to the formation of pseudomorphs made up of an aggregate of calcite crystals. The carbonation progressed from outside to the center of the gypsum crystal and occurred through a series of reactions, which involved the dissolution of gypsum and the concurrent crystallization of different polymorphs of $CaCO_3$ [amorphous calcium carbonate (ACC), vaterite, aragonite, and calcite], in addition to several solvent-mediated transformations among these polymorphs (Fig. 1.16). The degree of accuracy of the pseudomorph morphology obtained was correlated to the kinetics of the carbonation process, which in turn was determined by the original concentration of carbonate in the aqueous solutions. The conversion of gypsum into $CaCO_3$ through reaction with carbonate-bearing aqueous solutions is particularly complex as it encompasses the formation of one amorphous (ACC) and three crystalline polymorphs of $CaCO_3$ (vaterite, aragonite, and calcite). In the experiments presented in this study (Fernandez-Diaz et al., 2009), the dissolution–crystallization processes might be described as a sequence of solvent-mediated transformations and were the unavoidable result of the thermodynamic drives toward reducing the free energy of the system in the most efficient manner, i.e., at the maximum rate. Once a gypsum crystal is in contact with a carbonate solution, it starts to dissolve and, as a consequence, Ca^{2+} and SO_4^{2-} ions are released into the solution. Therefore, the solution quickly becomes supersaturated regarding all $CaCO_3$ solid phases cited above, at least in the neighborhood of the interface between the aqueous solution and the gypsum surface. Even though calcite is the stable polymorph of $CaCO_3$ at room temperature and the supersaturation concerning this phase will always be $\sim 300 \times$ higher than for ACC, in all the experiments the ACC is the first phase to form. As indicated by Sawada (1998), the creation of ACC is promoted by kinetic factors. This behavior is consistent with the predictions of the Ostwald Law of Stages stating that the nucleation of the most soluble and disordered phases is kinetically favored. Nucleation of ACC occurs straight on the surface of the gypsum crystal instead of in the aqueous solution. This could be as a result of both the substantial reduction of the activation energy that nucleation on a substrate embodies in comparison to homogeneous nucleation and the fact that supersaturation at the interface between the gypsum crystal and the solution is probably higher than in the aqueous solution. Nevertheless, a solution in equilibrium with ACC is still supersaturated in regard to other less soluble $CaCO_3$ polymorphs. This state constitutes the driving force for the coincident dissolution of ACC and the nucleation and growth of other $CaCO_3$ phases. This dissolution–crystallization process progresses concurrently with the progress of gypsum dissolution and begins shortly after the start of the experiments, only finishing when the ACC layer has totally

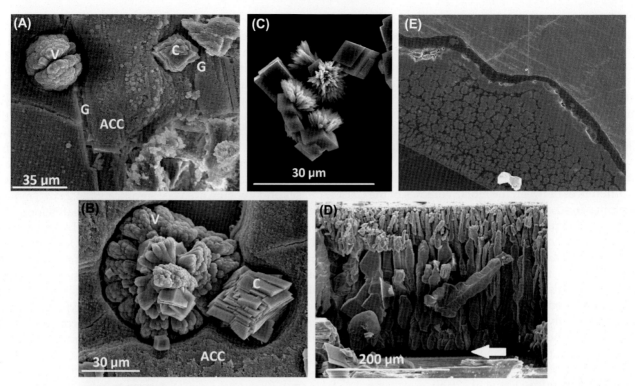

FIGURE 1.16 Scanning electron microscopy (SEM) micrographs showing the replacement of gypsum (G) and the amorphous calcium carbonate layer (ACC) by vaterite (V) and calcite (C). (A) The image has been taken of the original gypsum surface after a 45 min period in contact with a 0.5 M Na$_2$CO$_3$ aqueous solution. (B) The image corresponds to a reaction period of 60 min with a 0.5 M Na$_2$CO$_3$ aqueous solution. Note the growth of a calcite crystal against the surface of a vaterite spherulite. (C) Calcite rhombohedra and aragonite spherulites showing the typical fibrous-radial morphology, formed after a reaction period of 15 min. The concentration of the Na$_2$CO$_3$ aqueous solution was 0.05 M. (D) SEM micrographs showing the characteristics of the transformed layer, with the columns formed by calcite rhombohedra orientated perpendicular to the inward-moving gypsum surface. After 300 min interaction with a 0.25 M Na$_2$CO$_3$ aqueous solution. *Arrow* points to the small gap between the gypsum and the CaCO$_3$ layer. (E) SEM micrographs of a polished section of a gypsum crystal partially replaced by CaCO$_3$. Detail showing the porosity of the CaCO$_3$ layer and the small gap developed in between this layer and the inward moving surface of the gypsum crystal (Fernandez-Diaz et al., 2009).

disappeared. The various crystallization series detected, subject to the initial carbonate concentration in the solution, can be correlated with differences in the supersaturation level for the polymorphs of CaCO$_3$ during the early stages of the transformation processes. It is apparent that at the early stages, supersaturation is higher when more concentrated carbonate solutions are used. Consequently, when the carbonate concentration is high, the progress of gypsum dissolution preserves a high supersaturation with respect to other CaCO$_3$ phases even after the formation of the ACC layer. This helps the nucleation of metastable vaterite. Lastly, a further reduction of supersaturation ensuing the growth of vaterite favors the formation of the stable phase calcite. In contrast, when less concentrated solutions are used, the supersaturation decrease, a result of the formation of the ACC layer, readily results in the nucleation of calcite. Although the differences in the mineralogical order observed because of the starting carbonate concentration are obvious, in all the experiments several dissolution—crystallization loops, typical of solvent-mediated transformations, are concurrently operating during the carbonation process. Vaterite and/or aragonite crystals often form after the nucleation of calcite when low carbonate concentrations are used. This appears to suggest that the very high SO$_4^{2-}$/CO$_3^{2-}$ ratios in the solution help to stimulate the crystallization of both vaterite and aragonite with regard to calcite. The formation of aragonite and/or vaterite after calcite nucleation at all times happens very soon (reaction time ~15 min), when the carpeting of the gypsum surface is still very restricted. The observation is in line with the results reported by Bischof and Fyfe (1968) and Bischof (1968), who found that SO$_4^{2-}$ ions promoted the formation of aragonite with respect to calcite and have an impeding or hindering effect on the transformation of vaterite into aragonite and of aragonite into calcite via the solution. The textural features of the transformed layer are in line with the advance of a carbonation front into the gypsum crystals. There is predominantly strong proof of this in the orientation of the columns of calcite rhombohedra in the transformed layer. Such columns can be recognized as soon as calcite becomes the prevailing phase in the replacement and, in them, the rhombohedra display their three-fold axis oriented at right angles to the original surface of the reactant crystal. Furthermore, these calcite crystals exhibit marks of dendritic growth pointing to the inward-moving surface of the gypsum crystal. Hence, it suggests that the carbonation may be described as a process of mass transfer through an interfacial fluid film from the gypsum surface to the $CaCO_3$ layer. It is obvious that the thickening of the transformed layer strongly contributes to the slowdown of the carbonation progress. Besides, the armoring effect of the freshly formed layer becomes more significant as the process progresses. Nevertheless, the preservation of the initial volume of the gypsum crystal during the process necessitates the formation of microporosity. This is a common behavior when either the molar volume or the solubility of the product or both are smaller than those of the reactant phase. In the case under consideration, the differences between the molar volumes of the $CaCO_3$ phases and gypsum are very significant (34.2 cm^3/mol for aragonite, 36.40 cm^3/mol for calcite, 56.5 cm^3/mol for vaterite, and 74.5 cm^3/mol for gypsum). What is more, the solubility of gypsum is much higher than the solubility of any of the $CaCO_3$ phases (around 1000× higher than calcite's). Consequently, the amount of gypsum dissolved is much more than the amount of $CaCO_3$ precipitated (Putnis et al., 2005). Consequently, the volume reduction cannot be only balanced by the generation of microporosity. This situation is resolved by the growth of the observed gap between the inward-moving gypsum surface and the $CaCO_3$ layer. This gap is occupied by the interfacial liquid, which interconnects with the surrounding solution through the newly formed microporosity. It is worth indicating that the gap between the $CaCO_3$ layer and the gypsum surface, and consequently the volume of the interfacial liquid, grows as the carbonation progresses. This is a consequence of the fact that as the transformation progresses, the difference between the volume of gypsum that has dissolved and the volume of solid newly formed increases, i.e., the volume reduction that needs to be compensated increases (Fernandez-Diaz et al., 2009).

Solid-fluid interactions frequently comprise the replacement of one phase by another despite the fact that the morphology is retained as well as structural details of the parent phase, that is, pseudomorphism. In a study by Putnis et al. (2005) they reported in situ observations of the progression of both the solid and fluid compositions at the interface in the model system KBr–KCl–H_2O, in which a single crystal of KBr was replaced by a single crystal of KCl. The pseudomorphism was instigated by epitaxial growth at the fluid-mineral interface, after the dissolution of the parent phase resulted in an interfacial fluid layer that was supersaturated with regard to a different solid composition (Fig. 1.17). The following evolution of the coupled dissolution and growth can be related to local equilibria defined by a Lippmann diagram. The reaction characteristics, comprising the formation of porosity in the new solid phase, share many features of replacement reactions in nature as well as in technical applications. On contact with the KCl solution, the KBr crystal turned milky at an interface, which moved from the surface to the core of the crystal. The early formed rim was a Br-rich K(Br,Cl) phase and gradually became increasingly Cl-rich until the new crystal was in equilibrium with the new fluid composition. The degree of reaction is governed by the original solid and fluid volumes, but in the experiments described in this study, the final equilibrium composition reached after 2 h is almost pure KCl. Imaging by SEM and chemical analysis by EDX proved that the interface between parent

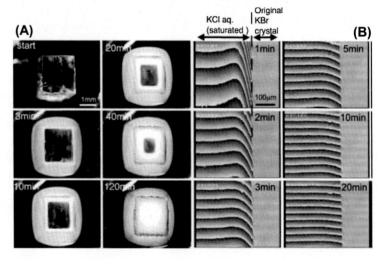

FIGURE 1.17   (A) A sequence of optical micrographs showing the migration of the replacement interface through the crystal. The initial KBr crystal is optically clear. As the reaction proceeds, the product, K(Br,Cl), appears milky, due to the development of þ ne porosity. To view the movement of the replacement interface, the fluid was added up to, but not covering, the top surface of the original KBr crystal. The resulting meniscus is seen as the white area surrounding the crystal. After 120 min the KBr crystal is completely replaced by a KCl crystal maintaining the same external volume. (B) A sequence of images from a real time phase shift interferometer. Parallel fringes are produced by a reference beam, which is inclined to the sample beam. The fringes represent characteristic refractive index values of the fluid. Changes in refractive index indicate compositional changes in the fluid. The curved interference fringes near the surface of the crystal show an initial steep compositional gradient at the surface, produced by the initial replacement reaction. The fringes only show changes outside the crystal surface. The reaction interface moves within the crystal, but the original external surface of the KBr crystal is preserved throughout the process (Putnis et al., 2005).

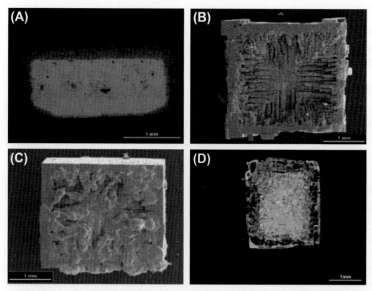

FIGURE 1.18 Images of cross-sections through replaced crystals. (A) Scanning electron microscopy (SEM) back-scattered image of the replacement interface in a crystal of KBr exposed to a saturated KCl solution for 10 min. The lighter center is the original KBr. The darker rim shows the porous replacement K(Br,Cl) product. (B) SEM image of a completely replaced crystal after 2 h reaction time. The composition is now KCl and the crystal has a coarse internal porosity developed into channels oriented toward the center of the crystal. The outside rim is no longer porous. (C) SEM image of the replaced crystal after 24 h, showing increased porosity coarsening. (D) Transmitted light microscope image of a replaced crystal after 12 days showing further coarsening, resulting in a transparent KCl crystal with a thicker nonporous rim. All scale bars are 1 mm (Putnis et al., 2005).

and product was sharp and the chemical compositions determined from line profiles were consistent with those determined from the lattice parameters (Fig. 1.18). Although the collective core and replacement rim diffracts like a single crystal (Putnis and Mezger, 2004), the rim was highly porous. This porosity was larger than that resulting from the molar volume reduction in replacing KBr by a K(Br,Cl) solid solution phase (Glikin et al., 1994; Pollok et al., 2011). As the reaction continued, the fine porosity, which instantly formed on initial reaction of KCl solution with KBr solid, started to first coarsen into channels and then, after conclusion of the chemical equilibration of the replacement, to close from the outside of the crystal, resulting in an almost porous free rim, enclosing residual fluid. At this stage the crystal appears transparent due to the absence of any fine porosity. Instantaneously after the saturated KCl solution came in contact with the KBr crystal, the Refractive Index (RI) of the solution at the crystal surface sharply increased. The abruptness of the RI gradient, i.e., concentration gradient, resulted from KBr dissolution generating an interface fluid composition rich in KBr. This also showed that the dissolution reaction took place faster than the fluid phase could equilibrate by diffusion. The fluid composition at the interface was straightaway supersaturated in relation to a Br-rich K(Br,Cl) solid solution phase that precipitated epitaxially (and therefore with crystallographic continuity) at identical rate as the dissolution of the parent KBr crystal. Interferometry, as well as the optical microscopy recording, showed that the position of the external surface of the original crystal stayed in place, and therefore there was no change in morphology or dimensions during the replacement process. This is typical of pseudomorphism. The experimental results point to a coupled dissolution—reprecipitation replacement mechanism controlled by the initial composition of a fluid boundary layer formed by dissolution of the original KBr crystal, and the simultaneous epitaxial precipitation of the product. The fluid—solid equilibrium is only preserved within this boundary layer whose width depends on the relative dissolution rate of the original KBr crystal, the epitaxial nucleation of the product crystal and the mass transport rate through the fluid, both to the crystal surface and within the porous replacement up to the reaction interface. This mechanism of pseudomorphism does not necessitate stress to maintain the external dimensions of the crystal. The transfer of crystallographic data from parent to product depends on the amount of lattice mismatch allowing for possible epitaxial precipitation, but in solid solution—aqueous solution systems where constant changes in lattice parameters go with the incremental changes in equilibrium compositions, a single crystal of the parent can be totally replaced by a single crystal of the product. The formation of porosity is a crucial feature of the replacement mechanism, letting fluid access the reaction interface and therefore the mass transport necessary for the replacement to proceed further within the crystal. The coarsening of the porosity is in line with reaching textural equilibrium forced by the decrease in surface area. Crack healing, basically the closing of porosity within a single crystal, is an analogous phenomenon, also driven by reduction of interfacial area (Brantley et al., 1990). As soon as the outer porosity closes, the internal porosity may be trapped and result in the formation of fluid inclusion trails. Once chemical equilibration is finished, continued textural equilibration over longer times proves that this porosity is a temporary phenomenon. The extent of porosity created hinges not only on the molar volumes of the parent and product crystals, but also on their relative solubility. Allowing for only molar volume changes, the estimated molar volume reduction in these experiments would be 13.2%. Nevertheless, bearing in mind the solubility differences, as well as molar volume changes, one would anticipate a total porosity of around 30% (Pollok et al., 2011). This elucidation uses both the KBr—KCl solubility diagram and the Lippmann diagram in which the equilibrium compositions suggest that the amount of solid dissolved must be larger than the amount of product precipitated, i.e., a volume deficit reaction. The interfacial control, which decrees that the reaction interface is the original crystal surface, unavoidably leads to the requirement of porosity in the product. In natural pseudomorphic replacements, the porosity can be a transient phenomenon, and textural reequilibration as observed also in these experiments, may result in a porosity-free product (Putnis et al., 2005).

Hydrothermal experiments of diopside interaction with $(Ni,Mg)Cl_2$ aqueous solutions were carried out by Majumdar et al. (2014) to elucidate the replacement mechanism and pattern of element mobilization and its significance to peridotite alteration. The diopside-pseudomorph interface is compositionally and structurally sharp. After the $MgCl_2$ experiments pure talc and/or serpentine were formed as the replacement phases, whereas reaction in the $NiCl_2$ fluids formed talc−willemseite and/or lizardite−nepouite series compositions. In addition, a complex rim of Ni-poor and Ni-rich regions developed in the $NiCl_2$ and $(Ni,Mg)Cl_2$ experiments (Fig. 1.19). Overgrowths of an olivine-type phase at 500−600°C and serpentine-type phases in 300−400°C experiments were also found. For a simple binary composition (Ni-Mg) of the reacting fluid, the fluid composition (different Ni/Mg ratio) at the reacting interface and its silica activity have a solid control on the phase formed. The consecutive formation of a Ni-poor inner phase and Ni-rich outer phase suggests a stepwise increase in Ni accommodation within the new-formed phase, which is illustrative of a cyclic dissolution−precipitation process in the course of supergene enrichment of Ni in natural ore deposits. In all the experiments, the diopside was incompletely replaced by reaction rims of variable thickness and composition, and consisted of talc- and serpentine-type phases fitting in either the talc $[Mg_3Si_4O_{10}(OH)_2]$−willemseite $[Ni_3Si_4O_{10}(OH)_2]$ or lizardite $[Mg_3Si_2O_5(OH)_4]$−nepouite $[Ni_3Si_2O_5(OH)_4]$ solid solution series. The reaction rims pseudomorphically substituted the diopside and in the reactions with Ni in the fluid phase, consisted of a double rim with a Ni-poor inner rim and a Ni-rich outer rim. In the experiments using pure $MgCl_2$ solution the newly formed phase was talc. In addition to the replacement phases, the outer rim was also often fringed with isolated crystals of olivine-type phases at higher temperatures (500−600°C) and a serpentine-type phases at lower temperatures (300−400°C). Reactions in pure $NiCl_2$ solution showed a partially replaced diopside grain after a reaction time of 15 days at 600°C. Although the reaction rim thickness varied across a single grain (from a maximum of 120 µm to a minimum of 20 µm), the inner low-Ni rim was always broader than the outer high-Ni phase, with the exception of the $NiCl_2$ experiment conducted at 300°C. The outer rim had a more constant width, which only ranged from 10 to 15 µm in width. There was no relationship between rim thickness and experiment temperature and different grains from the same experiment exhibited variable rim thicknesses. The interface between diopside and the Ni-poor phase is sharper than that between the Ni-poor and Ni-rich phases (Ni-rich talc and Mg-poor willemseite, respectively). A change in the replacement phase morphology was also seen between experiments at 500−600°C and 300−400°C. At higher temperatures, both Ni-poor and Ni-rich phases showed the same lamellar crystal habit, whereas at low temperatures the outer rim was made up of numerous densely packed small grains. One additional experiment using a pure $NiCl_2$ solution was carried out for only 6 days at 600°C and under these conditions only a single rim formed with a lamellar habit and composition very similar to the low-Ni rim. Isolated crystals of alternative Ni-enriched phase that enclosed the uninterrupted rim in 500−600°C experiments were identified as Mg-rich Ni olivine. Reactions in $(Ni,Mg)Cl_2$ solution in the higher temperature experiment, both a double rim of Ni-poor talc formed at the interface with the diopside together with an outer Ni-rich talc rim. An extra light-colored phase in the BSE image was identified as a Ni-rich Mg-olivine $[Mg_{1.43}Ni_{0.57}Si_2O_4]$. In the 400°C experiment, the inner phase was analyzed as a Ni-poor talc, but the outer phase fitted much better to a serpentine stoichiometry (Mg-rich nepouite). The same inner Ni-poor talc phase and outer Mg-rich nepouite rim were also observed in the 500°C and 300°C experiments. In the pure $MgCl_2$ solution experiments, only one uninterrupted replacement rim was created. The rim thickness was also much larger than those formed during experiments with the $NiCl_2$ solutions. The replaced phase exhibited a comparable texture (lamellar habit) as seen for the inner Ni-poor phase in Ni-bearing solutions. The phase precipitated in pure $MgCl_2$ solution experiments (at 600°C) was identified as the continuous pseudomorph rim of talc. In the high temperature experiments, an additional secondary mineral phase was precipitated over this continuous rim as isolated individual crystals that were identified as Mg olivine $[Mg_{2.000}Fe_{0.001}Si_{0.996}O_4]$. For 500°C experiments, the analytical totals from microprobe analysis of replacement phases were much lower (≈ 70 wt%) than those for ideal talc (92 wt%) or serpentine (87 wt%), which was ascribed to micro- to nanoporosity linked with the formed sheet silicates. Nevertheless, the Mg/Si (individual atom concentration is in wt%) ratio indicated a composition closer to serpentine. For the 400°C experiments with the same $MgCl_2$ solution, as well as the formation of a continuous inner talc rim, an outer phase of tightly packed grains was recognized at some isolated regions. These were comparable to those found in the outer Ni-rich rim in Ni-bearing solutions under similar temperature conditions and were closer to a serpentine stoichiometry. In the 300°C experiment using $MgCl_2$ solution, a similar shortage in the analytical totals of microprobe data (≈ 65 wt%) was observed and the Mg/Si ratio indicated a

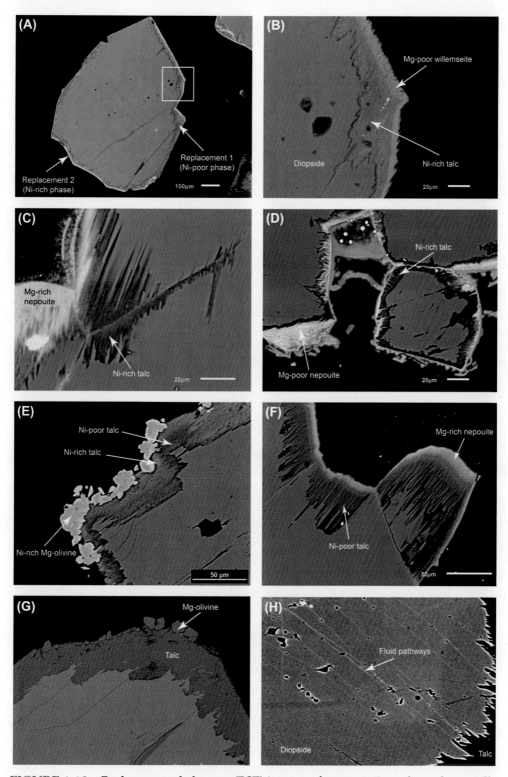

FIGURE 1.19 Back-scattered electron (BSE) images of cross-sections through partially reacted diopside grains. (A) Typical diopside grain after 15 days of experiment in 1 M $NiCl_2$ solution at 600°C and 100 MPa pressure showing pseudomorphic Ni-bearing phases. (B) Represents the area marked by *white box* in (A), which is a higher magnification image of the same diopside grain in (A). (C) Typical diopside grain close to the rim with the same $NiCl_2$ solution at 400°C. (D) A diopside–pseudomorph interface after reaction with the same solution at 300°C. (E) Details of a diopside crystal after pseudomorphism in an experiment with 1 M $Ni_{0.5}Mg_{0.5}Cl_2$ solution at 600°C and 100 MPa pressure for 15 days. (F) Partially pseudomorphed diopside grain in an experiment with 1 M $Ni_{0.5}Mg_{0.5}Cl_2$ solution at 400°C and 100 M Pa pressure for 15 days. (G) Lamellar talc after replacement of diopside in an experiment with $MgCl_2$ solution at 600°C. (H) Arrays of parallel cracks into the diopside close to the rim, emanating from a wedge-shaped area of the pseudomorphing talc phase toward the diopside crystal (Majumdar et al., 2014).

mixture of talc and serpentine. The replacement rim was in addition very thin ($\approx 10\,\mu m$). A further chemical analysis of that replacement rim showed a lower concentration of Mg than Si toward the diopside interface of the rim that gradually changed to a similar Mg/Si ratio toward the outer edge. Hence, serpentine might also be present over inner talc as a thin outer layer, but this was beyond the spatial resolution of the analytical techniques used in this study. Cracks filled with the replacement phases could also be found penetrating into the diopside crystal close to the rim after the experiments. A single rim of inner Ni-rich talc composition was formed in the 6-day experiment at 600°C with a NiCl$_2$ solution. Consequently, diopside to inner talc—willemseite series phase replacement is the initial stage of the bulk-replacement process and after that a replacement of the Ni poor by a Ni-rich phase. The inner talc—willemseite series phases at all times have higher concentrations of Mg than Ni. Isocon analyses (Grant, 1986, 2005) of diopside to inner talc—willemseite series phase replacement indicate that Mg and Si are relatively the two most conservative elements during the replacement. Thus diopside to inner low-Ni phase replacement reaction, for example, at 600°C with NiCl$_2$ solution, can be written as follows:

$$2CaMgSi_2O_6 + 0.9NiCl_2 + 2H_2O \rightarrow Mg_2Ni_{0.9}Si_4O_{10}(OH)_2 + 2Ca^{2+} + 2OH^- + 1.8Cl^- \tag{1.8}$$

Diopside → Ni-rich talc

This reaction is both Si and Mg conservative. The reaction also conserves solid volume as 1 mol of Ni-rich talc (molar volume $\approx 223.0\,\text{Å}^3$) replaced 2 mol of diopside (molar volume $\approx 110.2\,\text{Å}^3$). Therefore, diopside to Ni-rich talc replacement is also close enough to being isovolumetric assuming that there will also be some porosity generation accompanying the reaction (Putnis, 2009b; Putnis and Austrheim, 2013). Ni-rich talc to Mg-poor willemseite replacement at 600°C can be written as an isovolumetric (discounting volume change due to incorporation of different proportions of Ni) as well as Si conservative reaction:

$$Mg_{2.0}Ni_{0.9}Si_4O_{10}(OH)_2 + 1:5NiCl_2 \rightarrow Mg_{0.5}Ni_{2.4}Si_4O_{10}(OH)_2 + 1.5Mg^{2+} + 3Cl^- \tag{1.9}$$

Ni-rich talc → Mg-poor willemseite

In contrast, the Ni-rich talc to Mg-poor nepouite replacement at 300°C is neither Si nor Mg conservative, as indicated by isocon analyses. However, the formation of a dense Ni-enriched outer rim with minimal porosity due to small solid volume decrease (talc $\approx 223.0\,\text{Å}^3$, nepouite $\approx 177.5\,\text{Å}^3$) is closer to an isovolumetric reaction.

$$Mg_{1.4}Ni_{1.4}Si_4O_{10}(OH)_2 + 1.7NiCl_2 + 2H_2O \rightarrow 1.2Mg_{0.3}Ni_{2.6}Si_2O_5(OH)_4 + 1.6SiO_2aq + Mg^{2+} + 3.4Cl^- + 1.2H^+ \tag{1.10}$$

Ni-rich talc → Mg-poor nepouite

The isocon analysis of the diopside to pure talc replacement reaction suggests that the reaction may be Si conservative. Nevertheless, a broad inspection of the possible reactions suggests that the Si conservative reaction should also conserve the solid volume (diopside $\approx 110.2\,\text{Å}^3$ and talc $\approx 223.0\,\text{Å}^3$) via the following reaction.

$$2CaMgSi_2O_6 + MgCl_2 + 2H_2O \rightarrow Mg_3Si_4O_{10}(OH)_2 + 2Ca^{2+} + 2Cl^- + 2OH^-$$

Diopside → talc

The control of surface reactions on oxygen isotope exchange behavior during mineral—fluid interactions has been extensively studied (Cole and Chakraborty, 2001) and comprises processes such as dissolution—precipitation (Cole, 2000) and grain boundary diffusion (Cole et al., 2004). In the experiments in this study, The Si-O(bridging)-Si band in the Raman spectrum of the Ni-rich talc formed in an ^{18}O-enriched fluid was shifted by 10 ± 1 cm^{-1} toward lower wavenumbers relative to unspiked solution experiments, indicating that ^{18}O atoms were integrated into the Ni-phase silicate framework. Such ^{18}O incorporation has been found in a wide selection of mineral replacement reactions within a range of reaction conditions (e.g., leucite to analcime (Putnis et al., 2007c); albite to K-feldspar (Niedermeier et al., 2009); oligoclase and labradorite to albite (Hövelmann et al., 2010); calcite to apatite (Kasioptas et al., 2011); olivine to amorphous silica (King et al., 2011) and is proof for mineral replacement via an interface-coupled dissolution—precipitation mechanism (Putnis, 2002, 2009b; Putnis and Austrheim, 2013). This mechanism of pseudomorph formation after diopside is reinforced by BSE imaging along the sharp but irregular interface that divides the diopside from the replacing phase, exhibiting a large change in the BSE contrast and crystal habit across the interface. An analogous pattern for the replacement was also observed in the fractures inferring that they act as fluid conduits exposing new surfaces to the solution, and as a result enabling the replacement reaction. The recrystallization front moves from the external surface through the crystal to the core via a mechanism of mineral replacement that furthermore permits the transport of aqueous species through the reaction rim. Analytical results also exhibited a sudden decrease in Ca concentration across the replacement interface with an opposite pattern of Ni and Mg variation, which is similar to a step profile instead of a gradational change. This is consistent with the mechanism suggested for the replacement of chain silicates at lower temperatures (Daval et al., 2009; Ruiz-Agudo et al., 2012). Throughout the diopside replacement reaction a spatial coupling ensued between diopside dissolution and secondary Ni-bearing phase precipitation because a pseudomorph was formed in all the experiments. Consequently, dissolution of diopside must be the rate-limiting step under these conditions. In contrast, at 300°C when decoupling happened at the direct interface and precipitation occurred at a larger spatial distance from the dissolving diopside to create the gap (up to 10 μm) at the interface, diffusion of ions through the rim or formation of product phases can become the rate-limiting step of the reaction. The experiments in this study proved that, during diopside-Ni-bearing fluid interaction, Mg-rich talc-type phases are originally produced. A continued interaction of these replaced talc-type mineral phases with Ni-enriched fluids, concomitant with constant mobilization of Ni, would result in the precipitation of talc—willemseite or lizardite—nepouite series minerals as a still more Ni-rich phase. These two different stages of replacement, every one controlled via an interface-coupled dissolution-precipitation process, can be associated with two different natural processes respectively: (1) Diopside to inner Mg-rich/Ni-poor phase formation: The experiments offer evidence for Ca release during alteration of diopside where Ca transport is in the opposite direction to Ni and/or Mg. Austrheim and Prestvik (2008) have also observed the serpentinization of clinopyroxene at Leka ophiolite complex, Norway as the Ca-releasing reaction and ascribed this as the source of Ca for rodingitization. Therefore, Ca can be significant both as a geochemical tracer of economic Ni deposits over clinopyroxene-rich peridotite Lewis et al. (2006) and as source of Ca for the rodingitization processes during peridotite alteration. (2) Ni-poor phase to Ni-rich phase: The hydrothermal experiments proved that during inner Ni-poor to outer Ni-rich rim formation, the compositional Ni/Mg, Fe/Mg, and Ni/Fe ratio progressively increased. Throughout Ni enrichment in "hydrous Mg-silicate type" Ni-bearing minerals including garnierite (Freyssinet et al., 2005) under weathering conditions (Golightly, 1981), an analogous trend of Mg content variation against Ni and Fe is observed (Golightly and Arancibia, 1979; Villanova-de-Benavent et al., 2014). Here, Ni is understood as an imported element, whereas Fe is residual (Trescases, 1973), which is also appropriate for the system in this study as the starting bulk solution only has Ni or Mg. Second, textural—chemical associations between different Ni-bearing phases in natural Ni-ore deposits frequently exhibit the formation of compositionally homogeneous rims with distinct Ni content (Villanova-de-Benavent et al., 2014). This is in line with the experimental results at higher pressure temperature conditions where the inner as well as the outer rims were compositionally homogeneous. Finally, the results confirm that temperature merely controls the degree of spatial coupling between diopside dissolution and secondary phase precipitation. Therefore, there are strong parallels in textural—chemical relationships between different Ni-bearing talc- and serpentine-type phases found in natural Ni-ore deposits under weathering conditions and the hydrothermal experiments. Consequently, the formation of a serpentine-type phase over a talc-type composition or vice versa during Ni-enrichment (Villanova-de-Benavent et al., 2014) has to be a dissolution—precipitation controlled process knowing that pressure and temperature only control the kinetics of these reactions (Majumdar et al., 2014).

Pyrochlore-type ($A_2B_2O_6O'$) ceramics are being investigated for the immobilization of highly radioactive waste. Understanding the alteration process of such possible nuclear waste ceramic materials in aqueous solution is critical for the extrapolation of their long-term stability in a nuclear repository. Existing models on pyrochlore alteration are based on a diffusion-controlled hydration and ion exchange processes. Nevertheless, a study by Geisler et al. (2005a) present results of a hydrothermal experiment at 200°C with a natural, polycrystalline pyrochlore and ^{18}O-enriched aqueous solution, which are not compatible with a process based on solid-state diffusion. The data presented were in line with a pseudomorphic reaction that involved the dissolution of the pyrochlore starting material supplemented by the concurrent reprecipitation of a defect pyrochlore at a moving dissolution—reprecipitation front. TOF-SIMS (time-of-flight secondary ion mass spectrometry) and Raman spectroscopic data showed, for the first time, strong confirmation that the alteration of crystalline microlite pyrochlore was controlled by the dissolution and the coincident precipitation of a thermodynamically more stable, i.e., less soluble pyrochlore instead of by a diffusion-controlled leaching process (Fig. 1.20). Generally, such a replacement process is supposed to function at a moving front and preserves the shape of the parent phase, suggestive of a pseudomorphic reaction. Yet, no porosity was detected in the altered areas by transmission electron microscopy (TEM), probably reflecting the larger molar volume of the defect pyrochlore. This poses the question how fluid transport to the interface was sustained during the reaction. In earlier work on the replacement of a KBr crystal by KCl, it has been shown that porosity plays a crucial role throughout an interface-controlled dissolution—reprecipitation process (Putnis and Mezger, 2004), but was observed to seal once the whole KBr crystal was replaced by KCl (Putnis et al., 2005). It is, therefore, possible that any porosity in the altered zones was sealed as the reaction progresses, causing an increased control of the diffusion properties of the defect

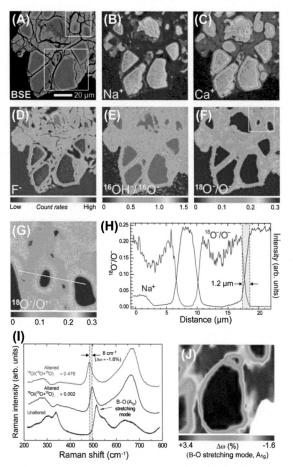

FIGURE 1.20 (A) Back-scattered electron (BSE) image of a polished aggregate of the run product showing altered areas (high BSE intensity) surrounding remnant unaltered regions (low BSE intensity). The large and small *white rectangles* outline the areas of the images shown in figures (G and J), respectively. (B–G) Time-of-flight secondary ion mass spectrometry (TOF-SIMS) secondary ion images of the same area shown in figure (A), which illustrate the distribution of Na^+, Ca^+, F^-, the $^{16}OH^-/^{16}O^-$ ratio, and the $^{18}O^-/(^{16}O^- + ^{18}O^-)$ ratio, respectively. Note that figure (G) shows a magnified section outlined by a *white rectangle* in figures (A) and (F). The secondary ion intensities or ratios are color coded on a linear scale. Each image is normalized to the most intense pixel shown in red. (H) Line scan of the $^{18}O^-/(^{16}O^- + ^{18}O^-)$ ratio and the Na^+ intensity along the profile that is marked by white line in figure (G). (I) Representative Raman spectra of the pyrochlore starting material and of altered areas from the experiments with a natural ^{16}O-rich and an ^{18}O-enriched solution. (J) Raman contour map of $\Delta\omega$ from a $54 \times 42\ \mu m^2$ large area, which has been created from 2268 Raman measurements. $\Delta\omega$ is the difference between the measured value of the B-O stretching frequency and the average frequency of the ^{16}O-rich altered areas ($=498\ cm^{-1}$). The mapped area is outlined by a *white rectangle* in figure (A) (Geisler et al., 2005a).

pyrochlore on the kinetics of the replacement process. As the reaction is probably controlled by the thermodynamics at the reaction interface, concentration variations within the fluid that moved along the grain boundaries might be responsible for the local enrichment of ^{18}O close to the interface (Geisler et al., 2005a). Pöml et al. (2007) exposed cuboids of a synthetic pyrochlore crystal and a natural metamict, i.e., self-irradion-damaged Ti-based pyrochlore to 1 mL of a 1 M HCl solution containing 43.5% ^{18}O at 250°C for 72 h. At the end of the experiments the colors of the samples had turned from black to nearly white and their surfaces were covered with micrometer-sized crystallites. Both samples were mostly altered into rutile with minor anatase and the reaction could be found several tens of micrometers into the cuboids. In the natural pyrochlore an extra phase (aeschynite $[(REE)(Ti,Nb)_2(O,OH)_2]$) was found as crystals on the surface and in an up to 100 μm thick layer at the interface between TiO_2 and the unreacted pyrochlore. The replacement reaction preserved even fine-scale morphological structures characteristic for pseudomorphs. This result accompanied by the enrichment of ^{18}O in the product phases and the textural features clearly demonstrate that the dissolution of pyrochlore is spatially coupled with the precipitation of stable (metastable) TiO_2 phases at a moving front. In the course of such a coupled process dissolution at the progressing front must be the rate controlling step. The aeschynite phase at the interface is believed to be either an intermediate alteration product or formed from a fluid near the interface between the TiO_2 polymorphs and the unreacted pyrochlore due to varying transport properties between the interface and the fluid with growing thickness of the TiO_2 rim. It has to be emphasized here that the results obtained under comparatively extreme batch-experimental conditions displayed resemblances with nature along with results ensuing from experiments conducted under moderate conditions such as expected in a nuclear repository. It is obvious that the rutile/anatase assemblage is highly enriched in Nb and U. Semiquantitative EDX analyses produced on average 14 wt% UO_2 and 41 wt% Nb_2O_5, as well as 5 wt% Ta_2O_5. The U concentration in TiO_2 is therefore considerably higher than in TiO_2 formed in earlier experiments under somewhat different conditions, even when allowing for the uncertainties related to EDX analyses. The high U content was attributed to micro- or nanometer-sized crystallites of U phases embedded in TiO_2, instead of being completely integrated in the rutile or anatase structure. In prior experiments with the same natural pyrochlore under analogous conditions, complex banding structures were observed, which were taken as being caused by a diffusion–reaction process, whereas the pyrochlore was in the solid state (Geisler et al., 2005b). It was inferred that in the course of the alteration reaction the REE, Y, and most of the U were selectively removed from the amorphous betafite, leaving the Ti, Nb, and Ta, permitting the nucleation and growth of anatase and the anatase-to-rutile conversion to take place. Nevertheless, the enrichment of ^{18}O in the reaction products and the textural features found in the current study clearly prove that the product phases crystallized from the solution. Therefore, both the self-irradiation damaged and the crystalline pyrochlore dissolved congruently in the HCl because the starting solution was undersaturated with regard to pyrochlore, but at that time the fluid–pyrochlore interface became instantaneously supersaturated in regard to TiO_2. The fact that the cuboids were still in one piece and the engraved letters on the surfaces were still present after the experiments suggests that the dissolution of pyrochlore is furthermore spatially coupled with the formation of the new product phases by preserving the original volume of the parent phase (Fig. 1.21). As argued by Putnis (2002) such a coupling of dissolution and precipitation may cause the complete replacement of one phase by an additional one without losing the external shape or crystal morphology of the initial phase as in pseudomorphism. The fact that the original volume is preserved during the reaction, even though the molar volumes and/or the solubilities of the parent and product phases are not the same, is a common observation in mineral replacement reactions (Labotka et al., 2004; Putnis and Mezger, 2004; Putnis et al., 2005). In preceding studies, for example, Geisler et al. (2004, 2005a,b) experimentally transformed a natural, crystalline Ta-based pyrochlore (microlite) under comparable conditions. The results demonstrated that the microlite was transformed to a defect pyrochlore that showed an epitactic crystallographic relationship with the initial pyrochlore. In the current case, though, the original phase is either polycrystalline or amorphous and there is no crystallographic relationship between the product and initial phases. Both rutile and anatase have a lower molar volume than radiation-damaged and crystalline Ti-based pyrochlore (rutile/anatase $\approx 20\ cm^3/mol$, betafite-(Y) $\approx 83\ cm^3/mol$, data from webmineral.com). In addition, in both situations the reaction comprises loss of elements to the fluid. It is obvious that the external volume can only be preserved through the creation of porosity in the reaction products. A crucial feature for the understanding of the replacement process in the radiation-damaged natural pyrochlore is the presence of a large gap running parallel to the surface and marking the boundary between the TiO_2 polymorphs and the unaltered pyrochlore.

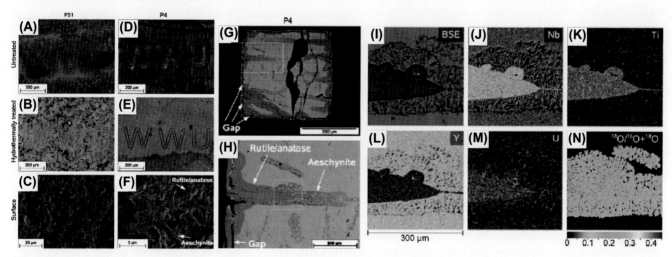

FIGURE 1.21 Secondary electron images of the surfaces of (A) and (D) the untreated samples PS1 (synthetic) and P4 (natural), respectively, and (B) and (E) the treated sample surfaces of both samples. The surface of the synthetic sample (B) changed dramatically. However, the letters "WWU" are still readable. Only little has changed on the surface of the natural pyrochlore (E). (C and F) Show detailed images of the PS1 and P4 surfaces. Relatively large crystals (up to 7 μm) grew on the surface of PS1, smaller crystals evolved on the P4 surface. Two different phases marked with *arrows* in (F) could be identified on the P4 surface. (G) Back-scattered electron (BSE) images of the cross-sections of the cuboid. The alteration of the natural P4 sample also took place along cracks. (H) The *rectangle* in (G) marks the areas shown in (H). The *rectangle* in (H) marks the areas mapped by time-of-flight secondary ion mass spectrometry (TOF-SIMS) and EDX (I–N). (I) BSE image of the areas analyzed by EDX and TOF-SIMS mapping in P4, (J–M) element distribution maps for Nb, Ti, Y, and U in the natural P4 sample. (N) Shows the $^{18}O/(^{16}O + ^{18}O)$ ratio for P4 (Pöml et al., 2007).

As the pyrochlore is nearly completely amorphous, the early nucleation of the Ta- and Nb-rich TiO_2 polymorphs onto the pyrochlore surface must have happened indiscriminately with no crystallographic relationship between the individual TiO_2 nuclei after a couple of pyrochlore surface layers dissolved. As the entropy of the anatase-to-rutile transformation is small (Schuiling and Vink, 1967), the crystallization of anatase almost certainly occurred metastably, because rutile is the thermodynamically stable TiO_2 polymorph in solutions within a large temperature range. This is consistent with the fact that nanocrystalline anatase is a common polymorph in synthetic and natural samples at low temperatures (Navrotsky, 2002). It is therefore probable that anatase initially nucleated and, with particle coarsening, anatase changed into rutile. It was pointed out that the original dissolution—reprecipitation step cannot have involved the total covering of the entire pyrochlore surface because of the difference in the molar volumes of the product and starting phases. Therefore, the dissolution of pyrochlore could progress as soon as the Ti concentration at the interface solution was reduced due to the nucleation/crystallization of TiO_2. It is obvious that the subsequent nucleation of TiO_2 ensued at the preexisting TiO_2 crystals instead of at the inward pervading new surface of the pyrochlore, because such an epitaxial nucleation process has to be energetically beneficial when compared with the heterogeneous nucleation on the amorphous pyrochlore surface. Because of the random crystallographic orientation of the originally formed crystals, the TiO_2 crystals form a polycrystalline layer that still preserves even fine morphological details of the original phase. The replacement process can be explained as a transport of material via the interfacial fluid film from the pyrochlore surface to the TiO_2 layer. Through such a process, the gap between the pyrochlore surface and the freshly formed TiO_2 layer and consequently the volume of the interfacial fluid increased. The large gap between the TiO_2 layer and the pyrochlore is therefore probably the relict of such a fluid film. As the growing TiO_2 layer also enclosed some porosity, the interfacial fluid was not confined, but could connect with the surrounding fluid. Because of the preexisting cracks in the cuboid of the natural sample, the fluid could also penetrate further into the cuboid. It is obvious that the replacement reaction spread out perpendicular to the direction of these cracks, which caused a local closing of the gap between the TiO_2 layer and the pyrochlore. The existence of the aeschynite phase both as thick layers inside the cuboid and as crystallites on the surface of the cuboid is not yet entirely understood. The phase could be assumed to be an intermediate alteration product. This would suggest that the pyrochlore originally recrystallized into aeschynite before it was successively transformed into rutile/anatase as the alteration process progressed. The aeschynite crystals on the surface of the sample may be relicts of such an intermediate alteration product. However, the favored explanation is that the aeschynite phase formed from an interfacial fluid between the unreacted pyrochlore and the alteration products rutile/anatase that was supersaturated in regard to aeschynite. This supersaturation could progress as the chemical exchange between the external and interfacial fluid slowed down because of an increasing penetration of the reaction into the cuboid and a growing thickness of the TiO_2 layer. If this was true, the presence of a thick layer of aeschynite crystals inside the cuboid is a direct proof for an increasing deviation from equilibrium between the external and the interfacial fluid. The results for the synthetic sample were marginally different due to its polycrystalline character. In this case, the fluid could migrate into the sample along the grain boundaries and through the open space provided by the primary porosity. No clear reaction front could be found on the large scale, but a close-up look showed numerous partially replaced pyrochlore grains that were attacked by the fluid migrating along the grain boundaries. TEM investigations were necessary to study the crystallographic connection between anatase/rutile and pyrochlore. Remarkably, the zone around the unaltered core was described by a considerably increased porosity in comparison to the primary porosity of the ceramic. This increased porosity suggests that the Ti was not immediately reprecipitated as TiO_2 inside the ceramic, but was first transported from this central core more to the outside. The cause for this is at this time unidentified. In comparison to the natural sample the reaction progressed deeper into the crystalline sample, because the ceramic was polycrystalline and the fluid might thus migrate into the sample along grain boundaries. Still, the grain size of the ceramic ranges from 8 to 60 μm in diameter. Even near the surface of the cuboid not all pyrochlore grains are totally replaced by TiO_2, which indicates that the reaction would most likely not have progressed deeper than around 10—30 μm for a single crystal (note that the grains were attacked from all sides). This would be below half the thickness of the replacement zone found in the natural radiation-damaged sample, suggesting that the dissolution of pyrochlore (and the reprecipitation of TiO_2) was faster in the radiation-damaged pyrochlore. This could also elucidate the result that the TiO_2 crystals at the surface of this sample are considerably smaller than those at the surface of the crystalline ceramic, because fast dissolution caused a high supersaturation at the interface and consequently in a high nucleation rate, i.e., there was less time for the TiO_2 crystals to grow before more nuclei formed (Pöml et al., 2007).

Oligoclase and labradorite crystals were experimentally replaced by albite in an aqueous sodium silicate solution at 600°C and 2 kbars in a study by Hövelmann et al. (2010). This replacement was pseudomorphic and was described by a sharp chemical interface that moves through the feldspar while preserving the crystallographic orientation. Reaction rims of albite, up to 50 μm thick, were obtained within 14 days. Reequilibration of plagioclase in an

^{18}O-enriched sodium- and silica-bearing solution resulted in oxygen isotope redistribution in the feldspar structure. The experimental properties of the reaction products are comparable to naturally albitized plagioclase and are symptomatic of an interface-coupled dissolution−reprecipitation mechanism. Chemical analyses proved that the albitization went together with by the mobilization of major, minor, and trace elements, also comprising elements such as Al and Ti that are usually considered as immobile during hydrothermal alteration. After the experimental treatment, the sample crystals transformed from optically clear to turbid. Nevertheless, the dimensions of the crystal grains did not change. SEM surface images showed a high porosity and roughness of the reacted material relative to the unreacted grains, which exhibited flat and smooth surfaces devoid of any visible porosity. Larger pores were often occupied with needle shaped crystals of pectolite ($NaCa_2Si_3O_8OH$). In nature, pectolite is a characteristic hydrothermal mineral that can be found in cavities and joint planes in basic igneous rocks and is common in metamorphosed high calcium rocks and skarns. Back-scattered secondary electron images and combined EDX analyses of cross-sections through the crystals point to pseudomorphic replacement of the original plagioclase by an albite-rich feldspar composition (Fig. 1.22). The replacement progressed from the grain surfaces and along preexisting cracks into the crystal, so that the product albite was located in a rim around the parent plagioclase and around the cracks. Nevertheless, the width of the reaction rim varied widely from grain to grain and even around a single grain, but there were no substantial differences of the average rim widths when comparing the 14-day experiments to the 21-day experiments. In some sections of partially replaced labradorite crystals, there were zones at the margin between the original labradorite and the product albite, which exhibited a dissimilar, more intricate texture with abundant large pores. BSE images with higher magnification indicated that these zones were consisting of an intergrowth between pectolite + albite and/or labradorite. The replacement product of both oligoclase and labradorite was almost pure albite ($Ab_{99-100}Or_{0-1}$), even though minor concentrations in K_2O or FeO could be identified in case of altered oligoclase or labradorite, respectively. Profiles of molar fractions of albite and anorthite proved that the major compositional change was sharp on a micrometer scale. Elemental distribution maps of Si, Al, Ca and Na showed that the albite reaction rims were chemically homogenous. It is evident that the product albite showed an extraordinarily different diffraction contrast compared to the starting plagioclase. Although the contrast was uniform and smooth in the unaltered part, the altered part showed an intricate, dense and dark structure indicating a much higher defect density. It is significant that these characteristics are comparable to those described from naturally albitized plagioclase. An assessment of electron diffraction patterns from the unaltered and altered parts as well as from the interface region showed that the crystallographic orientation of the oligoclase was maintained in the albite during the replacement process. All the experimental results suggest that the replacement of plagioclase by albite, i.e., the coupled exchange of $Ca^{2+} + Al^{3+}$ by $Na^+ + Si^{4+}$, progresses via an interface-coupled dissolution−reprecipitation mechanism. The chemical interface between the unaltered plagioclase and the product albite is sharp on the micrometer scale as shown in BSE images, in EMP profiles, and in elemental distribution maps. A solid-state diffusion process is not capable to create such a sharp chemical change along the replacement interface. In addition, Raman spectroscopy evidently proves that the oxygen isotopes in the framework structure were rearranged during reequilibration in the ^{18}O-enriched fluid. An approximation of the ^{18}O-content points to the product albite almost completely inheriting the oxygen isotope composition of the fluid. This is proof that structural Si−O and Al−O bonds have been broken during the replacement and that the albite formed from the fluid. The molar volume decrease from plagioclase to albite is only around 1% and can as a result only play an insignificant role in the formation of the porosity. Consequently, the porosity has to primarily originate from the solubility difference between the two feldspars in the fluid. As albite is more stable (or less soluble) than plagioclase in an aqueous sodium silicate solution, it follows that the amount of dissolving plagioclase is larger than the amount of reprecipitating albite. The observation that pectolite is often found in indentations at the crystal surface indicates that some of the surface porosity is associated with the growth of pectolite. Continuing albitization increases the Ca concentration in the fluid until it turns supersaturated with respect to pectolite. Pectolite can then begin to nucleate onto the surface of the partially replaced crystal. This can in the vicinity lower the degree of saturation with respect to albite, comparable to a flushing of the system, so that consequently some albite would dissolve, producing a notch at the surface. The unusual diffraction contrast that infers considerable strain, as well as the observation that the albitized parts are far more sensitive to the electron beam than the starting material, may point to a higher porosity that is in all probability present on the submicron scale. Once the reactivity of the fluid has decreased, it is furthermore conceivable that a part of the porosity was removed by textural reequilibration and recrystallization of pore space due to ripening processes, as has been witnessed experimentally in other replacement processes (Putnis et al., 2005). Ripening processes combined with pectolite precipitation, closing surface porosity, block a substantial part of the accessible fluid pathways and decelerate the forward movement of the reaction interface. Experiments by Harlov et al. (2005) showed that solid-state diffusion may become the principal process at the interface as a result of this effect and thus blurs the sharpness of the chemical interface. The observation that the average replacement rim widths after 14 and

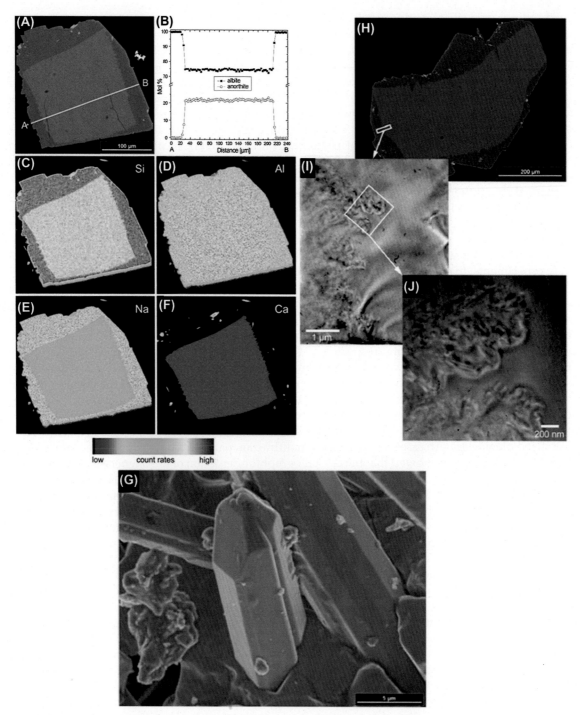

FIGURE 1.22   (A) Back-scattered electron (BSE) image of a partly albitized oligoclase grain that was chosen for an EMP line scan and element mapping. The direction of the line scan is indicated by the *white line*. (B) Profile of the molar concentration of the albite and anorthite component. Note the major compositional changes at the interface. The altered part of the crystal is chemically homogeneous and consists of nearly pure albite in comparison to the intermediate composition of the unaltered part. (C−F) Elemental distribution maps of Si, Al, Na and Ca, respectively. The sharp compositional interface and the lack of any chemical zoning within the altered part are again obvious. Note that areas showing a higher Ca-concentration within the albite rim correspond to the pectolite grains visible from the BSE image in (A). (G) Reaction products are rough with numerous pores which are often filled with needle shaped crystals of pectolite. (H) BSE image of a partly albitized oligoclase grain indicating the location where an focused ion beam (FIB) cut was performed (*white rectangle*). (I) Transmission electron microscopy (TEM) bright-field image of the sampled TEM foil. A detailed image of the area outlined by the *white rectangle* is shown in (J). The altered part (left side) shows a complex and dark diffraction contrast compared to the very uniform appearance of the unaltered part (right side) (Hövelmann et al., 2010).

21 days were not markedly different proves that the process of dissolution and reprecipitation was indeed slowed down. Nevertheless, as the interface was sharp on the nanoscale, as observed by TEM, solid-state diffusion is still secondary to dissolution and reprecipitation within the timeframe of these experiments. The fact that the albite reaction rim width differs from grain to grain and around/within single grains can have several reasons. It has been shown that Al—O bonds are energetically easier to break than Si—O bonds (Xiao and Lasaga, 1994). Therefore, it appears probable that the reaction kinetics are affected by the crystallographic orientation when considering the atomic sites exposed on different crystal surfaces (Hövelmann et al., 2010).

The mechanism of reequilibration of albite in a hydrothermal fluid was investigated experimentally using natural albite crystals in an aqueous KCl solution enriched in ^{18}O at 600°C and 2 kbar pressure by Niedermeier et al. (2009). The reaction was shown to be pseudomorphic and created a rim of K-feldspar with a sharp interface on a nanoscale which moved into the initial albite with increasing reaction time. The K-feldspar had a very high defect level and a disordered Al, Si distribution, in comparison to the original albite. The textural and chemical features in addition to the kinetics of the replacement of albite by K-feldspar were in line with an interface-coupled dissolution—reprecipitation mechanism (Putnis, 2002, 2009b; Putnis and Austrheim, 2010, 2013). After the experimental treatment, the cleaved feldspar grains did not change in habit, but the surfaces changed to a milky-white color and were fairly friable. The grains that were optically transparent before the reaction became translucent and some grains joined during the reaction to create clusters of various single crystals. The experimentally treated albite exhibited a high porosity at the grain surface, although there was no porosity in the initial albite. Some areas stick out from the mineral surfaces, whereas simultaneously; some pits penetrate deep into the albite crystal. The surface topography has enlarged on the reacted surface compared to the pristine albite surface, which only showed flat cleavage planes. The product K-feldspar exhibited a fine texture seemingly made up from small clusters, quite unlike the original albite surface. These observations indicate that new crystal growth was related to the reaction. Feldspar grains that were reacted for longer periods exhibited flat surfaces. At high magnification the surface of the product K-feldspar was shown to have a fine texture ostensibly consisting of small clusters, rather unlike the initial albite surface at the same magnification. In BSE images of sectioned grains a dark gray core and a brighter rim, representing the pristine albite and the reacted rim, respectively, can be recognized (Fig. 1.23). This change in image contrast point to a sudden change in the chemistry at the margin between the two phases. This margin or interface is sharp on the micrometer scale and is supplemented by a large number of pores. The thickness of the K-rich feldspar rim differed noticeably, e.g., from 10 to 100 μm in a 7-day experiment. EDX elemental mapping revealed that the converted rim principally contained K but also minor amounts of Na. Relics of pristine albite were unevenly dispersed within the grain. At the interface the feldspar composition sharply changed from pure albite to pure K-feldspar (Fig. 1.24). Remarkably, the pure K-feldspar merely occurred right near the interface in the reacted material. More away from the interface the formed K-feldspar became more Na-rich and approached an average composition $Or_{87}Ab_{12}An_1$. TEM in the bright field imaging mode showed a strongly different diffraction contrast between the original albite and the formed K-feldspar. The parent albite had smooth diffraction contours, indicating a rather defect-free crystal structure, whereas the product K-feldspar exhibited a complex diffraction contrast and numerous linear features, suggesting a high defect density. The product K-feldspar exhibited no macroscopic porosity, but dense dark textures near the boundary to the pristine albite. Some of these structures showed a stretched out form, tending into the direction of the fluid/solid interface, and might represent voids or channels. The interface between the two phases consisted of a dark zone that might be the result of a high density of dislocations. This dark area commonly appeared at the margin between the two feldspar phases. The crystallographic orientation of the K-feldspar rim corresponds within a degree with that of the albite core. Only a minor displacement was seen in the diffraction patterns of the interface area, with splitting between two comparable diffraction spots, from each feldspar phase. Some sections of the reaction interface exhibited dense dislocation arrays

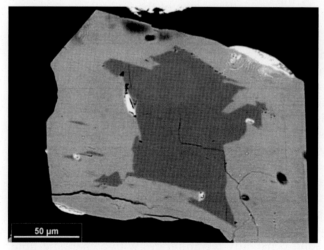

FIGURE 1.23 Scanning electron microscopy (SEM) image of a cross-section of the interface between pristine albite in the core and reacted K-feldspar in the rim. Note the occurrence of numerous pores at the interface (Niedermeier et al., 2009).

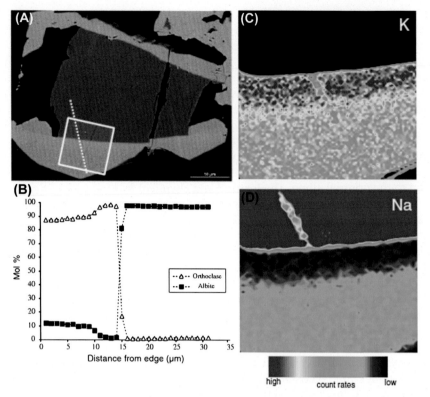

FIGURE 1.24   (A) Back-scattered electron (BSE) image of the area within a reacted feldspar grain that was chosen for electron microprobe traverse along the *white line* and for electron microprobe mapping marked by the *white rectangle*. (B) Profiles of the albite and orthoclase components across the interface of the product K-feldspar and unreacted albite. Note the decrease of the albite component before the reaction interface between 10 and 14 μm. Electron microprobe distribution maps of (C) K and (D) Na near the reaction front. Note that the interface between the pristine albite and the product K-feldspar is sharp on the micrometer scale. The exceptionally low Na values within the red area are caused by the line analyses (B), where Na and K were lost under the influence of the electron beam (Niedermeier et al., 2009).

indicating a semicoherent interface, whereas in others there seemed to be an open porosity. The results of this study indicate that a constant exchange by a coupled dissolution–reprecipitation process led to the transformation of albite to K-feldspar. The SEM observations of the reacted surface and the electron diffraction patterns across the reaction interface proved that, even though the K-feldspar had recrystallized throughout the reaction, it basically persisted as a single crystal with a similar crystallographic orientation as the initial phase. This is fully in line with an interface-coupled dissolution–reprecipitation mechanism in which the nucleation and growth of the product is epitaxial on the parent surface (Putnis and Putnis, 2007). When albite is substituted by K-feldspar, a molar volume increase of 8%, triggered by the incorporation of the larger K-cation, has to be compensated by the crystal lattice at the interface. It is the coherent stress that results in the formation of connected dislocations, which are responsible for pathways for chemical exchange between the surrounding fluid and the fluid at the reaction interface. After the experimental reaction larger open pores are common at the margin between the initial albite and the formed K-feldspar. These pores must have been created in the course of the hydrothermal experiments as they can only be found near the reaction interface and it is thus

highly probable that they are an essential feature of the replacement mechanism. It is obvious that the reaction rate should be influenced by accessible fluid pathways through the product crystal, i.e., on interconnected porosity. Kinetically, the reaction between albite and a KCl solution is highly reliant on the crystallographic direction, as shown by the irregular rim thickness. The reaction rate is higher along the (001) crystal surface, which exhibits more obvious structural alterations than other crystal surfaces. This can be understood based on the fact that the bonding parallel to the (001) lattice plane is weaker than along other crystal planes, which is revealed by the preferential cleavage of albite along the (001) plane. In ordered feldspars the percentage of Al-tetrahedra at the (001) surface is higher than at other crystal surfaces, and the Al-tetrahedra are favored sites for dissolution, because the Al–O bonds are energetically easier to break than the Si–O bonds. Volume diffusion is undeniably too slow as an effective transport mechanism of K and Na ions and O to explain the high exchange rates that were observed by Raman spectroscopy and electron microprobe. Nevertheless, on a small scale volume diffusion possibly might operate to equilibrate relict domains of albite, or to gradually homogenize areas within the reprecipitated K-feldspar. The specifics of the reaction progress by such a mechanism is probably intricate as the product solid constantly reequilibrates with a changing fluid composition. The approach to equilibrium conditions between the product feldspar and the composition of the fluid also undeniably influences the reaction rate. As the replacement of albite by K-feldspar progresses, the composition of the fluid alters to higher NaCl concentrations, whereas the product feldspar composition constantly reequilibrates through interaction with the fluid. At each step of such a progress, reequilibration is realized by consecutive dissolution and reprecipitation reactions. A further limitation though, is that such an explanation accepts that the rim composition is in equilibrium with the bulk fluid composition, although in situ experiments on replacement in salt systems have shown that the solid composition that reprecipitates is controlled by the fluid composition at the reaction interface, and that great compositional gradients can exist in the fluid as the reaction progresses (Putnis and Mezger, 2004; Putnis et al., 2005). This indicates that the rate limiting step in such replacements could be the mass transfer through the fluid pathways in the solid reaction product (Niedermeier et al., 2009).

In nature, ilmenite ($FeTiO_3$) weathers through oxidation and removal of Fe to form a seemingly continuous series of compositions from ilmenite to pseudorutile (theoretically $Fe_2Ti_3O_9$), and with continued weathering, to leucoxene. To understand the mechanism of acid-leaching of ilmenite to finally form rutile, Janssen et al. (2008, 2010) have carried out an experimental study of ilmenite alteration in autoclaves at 150°C in HCl solutions. The results indicate that the alteration may proceed in two stages, each with a sharp interface between the parent phase and the product. The alteration begins at the original ilmenite crystal surface and has also taken place along an intricate branching network of fractures in the ilmenite, generated by the reaction, through which the fluid can migrate. Element-distribution maps and chemical analyses of the reaction product in the fractures exhibit noticeable reduction in Fe and Mn and a relative enrichment of Ti. Chemical analyses though, do not match any stoichiometric composition, and may signify mixtures of TiO_2 and Fe_2O_3. The fracturing is conceivably driven by volume changes associated with dissolution of ilmenite and simultaneous reprecipitation of the product phases from an interfacial solution along an inward moving dissolution—reprecipitation front. The first alteration product formed was pseudorutile—no intermediate phases between ilmenite and pseudorutile were found. The textural relationship between ilmenite and pseudorutile points to a coupled dissolution—reprecipitation mechanism instead of a solid-state continuous oxidation and Fe diffusion mechanism. The second step involved a further dissolution—reprecipitation phase to form rutile. Raman spectroscopy revealed that the ^{18}O was integrated in the rutile structure during the recrystallization. During the whole alteration process, the original morphology of the ilmenite was preserved even though the product was very porous. The rutile inherited crystallographic info from the original ilmenite, producing a triply twinned rutile microstructure. The reaction can be explained in terms of an interface-coupled dissolution—precipitation mechanism. SEM images exhibited, after reaction, the initial surfaces covered with finely polycrystalline, closely spaced new crystals. It is obvious that with increasing acid concentration the surface structure altered drastically, with an increase in the size of the crystals and a more intensely developed morphology. After treatment with 0.5 M HCl, the crystals were approximately equidimensional, whereas at higher acid concentrations the tetragonal crystals were elongate and exhibited well developed twinning (Fig. 1.25). These rutile twins were the result of epitaxial growth on structurally related ilmenite crystals. The most remarkable result of this work is that the reaction from ilmenite to form rutile is pseudomorphic, i.e., preserves the external shape of the ilmenite notwithstanding the fact that big changes in chemical composition and solid molar volume are involved. In SEM two zones of noticeably different BSE intensity, a dark rim and a bright core indicate the degree of reaction. The bright core is the seemingly original unaltered ilmenite, whereas the darker rim is the replacement product rutile and exhibits a high porosity. The change from ilmenite to rutile produces a volume

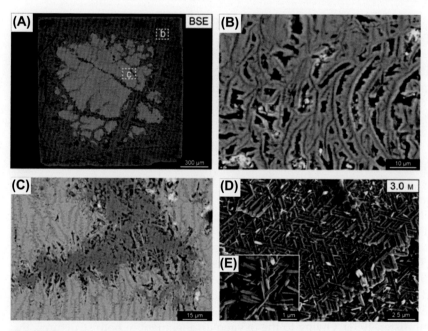

FIGURE 1.25 Back-scattered electron images of cross-sections through a partially reacted ilmenite cube at increasing magnification. (A) The whole ilmenite cube, showing the reaction rim of TiO_2 (darker gray) and the unreacted ilmenite core (light gray). The inset squares show the areas imaged in (B), the high porosity of the replacement product TiO_2, and (C) the sharp replacement interface between ilmenite and TiO_2. (D–E) in a 3 M HCl solution the surface is covered by prismatic rutile crystals in three orientations (Janssen et al., 2010).

reduction of about 40%. Electron microprobe analysis of the rutile rim yielded >90% TiO_2 with small amounts of FeO. In addition, the BSE images prove that the interface between the ilmenite and the alteration product rutile is sharp on the scale of micrometers. The replacement of ilmenite by rutile encompasses substantial changes in the molar volume because of the iron is entirely removed from the ilmenite. When substantial changes in molar volume take place between parent and product phases, along with differences in relative solubilities of parent and product phases, the product has to have porosity and/or cracks. This in turn provides pathways for mass transport through the fluid phase to the site of reaction. The reaction interface can hence migrate from the initial surface into the ilmenite. Moreover, even crystallographic information can be conveyed from the parent to the product (Putnis, 2009b). Pseudomorphism in the course of the replacement reaction may appear to indicate that the Fe is removed through "leaching". Usually, leaching is defined as the selective removal of some cations through an inert anion sublattice and the inference is that the mechanism comprises a solid-state diffusion process. The common observation that fine textural details along with crystallographic details of the parent phase are preserved in the product has been used to support such a mechanism. Nevertheless, Raman spectroscopy of samples where ^{18}O-enriched acid solution was used show that oxygen from the solution is assimilated in the replacement product rutile. This proves that the hexagonal closed-packed sublattice of ilmenite is totally reconstructed by the reaction, which is inconsistent with the hypothesis that a solid-state diffusion process controls the replacement of ilmenite by rutile. The textures together with ^{18}O results suggest that the replacement of ilmenite by rutile has all the features of an interface-coupled dissolution−reprecipitation mechanism as defined by Putnis (2002) and Putnis and Putnis (2007). Experimental work on hydrothermal alteration of pyrochlore has also proven that leaching involves a coupled dissolution−reprecipitation mechanism (Geisler et al., 2005a; Geisler et al., 2005b; Pöml et al., 2007). The preservation of external morphology, that is, formation of pseudomorphs, necessitates that the rate of dissolution of the ilmenite must be equal to the rate of precipitation of rutile, which is not an a priori clear condition for two different solid phases, and has to impose specific restrictions on the composition of the fluid at the reaction interface. The overall reaction may be given as:

$$FeTiO_3 + 2H^+ \rightarrow TiO_2 + Fe^{2+} + H_2O \tag{1.11}$$

or alternatively,

$$4FeTiO_3 + O_2 \rightarrow 4TiO_2 + 2Fe_2O_3 \tag{1.12}$$

when oxidation and precipitation of hematite is also taken into account. Using Eq. (1.11) for illustration purposes, the corresponding dissolution of ilmenite and precipitation of rutile can be written individually as Eqs. (1.13) and (1.14) below:

$$FeTiO_3 + 5H^+ \rightarrow Ti(OH)^{3+} + Fe^{2+} + 2H_2O \tag{1.13}$$

$$Ti(OH)^{3+} + H_2O \rightarrow TiO_2 + 3H^+ \tag{1.14}$$

where the $Ti(OH)^{3+}$ is the main species in solution at 150°C and 3 M HCl. These equations demonstrate that protons promote ilmenite dissolution but inhibit rutile precipitation, and consequently the overall rate of the reaction increases with increasing proton concentration (from 0.1 to 3 M) and ilmenite dissolution is the rate-limiting step. This backs the proposition by Xia et al. (2009a) and Zhao et al. (2009) that for effective coupling between dissolution and precipitation, dissolution has to be the rate-limiting step. The nucleation rate of rutile is therefore controlled by the fluid chemistry, i.e., by the supersaturation at the interface, but also by the epitaxial correlation between the two structures. When the crystallographic structures of the parent and product are comparable, i.e., a high degree of epitaxy, the nucleation rate on the surface of the parent phase will be boosted (Putnis and Putnis, 2007). In naturally altered rocks, rutile is frequently found epitaxial on ilmenite (Armbruster, 1981). Owing to the oxygen arrangement in both structures, the nucleation of rutile from the interfacial solution may be able to occur in three symmetrically equivalent orientations. This is observed as rutile prisms in a triply twinned orientation. This is a result of the epitaxial nucleation of rutile on the ilmenite surface, whereby the preservation of the roughly hexagonal close-packed oxygen and the symmetry change from hexagonal to tetragonal would be anticipated to result in three symmetrically equivalent orientations. An important observation in replacement reactions, involving molar volume changes, is the formation of fracturing, which is often induced by the reaction. The creation of cracks increases the penetration of the fluid toward the reaction front (Malthe-Sørenssen et al., 2006) and guarantees a constant flow of the fluid phase to and from the reaction front. In addition, the fragmentation process creates fresh surfaces within the parent phase, which can react quickly with the surrounding fluid (Jamtveit et al., 2009) and consequently increases the alteration process. The finger-like reaction structures must have been created through the fragmentation and as a result the rate of fracture generation controls the rate of the replacement process (Janssen et al., 2010).

The synthetic Mn oxide minerals todorokite and birnessite when exposed to Fe^{2+} ions produced a pseudomorphic precipitation of Fe oxide minerals. The pseudomorphism was a macroscopic phenomenon; the Fe oxide precipitate comprised fine crystallites of feroxyhyte, lepidocrocite, or akaganeite in an unattached network maintaining the form of the original Mn oxide mineral. The anions present and the original structure of the Mn oxide mineral affected the nature of the Fe oxide precipitated. Reactions of $FeSO_4$ (0.05 M) with Na-birnessite, todorokite, or soil minerals having birnessite and lithiophorite produced sulfate-adsorbed feroxyhyte as the major reaction product (Fig. 1.26). Reactions of $Fe(ClO_4)_2 \cdot 6H_2O$ (0.05 M) with Na-birnessite generated lepidocrocite and with todorokite produced lepidocrocite and akaganeite. Reactions of $FeCl_2$ with birnessite or todorokite resulted in akaganeite as the major phase. All these Fe oxides precipitated were poorly crystalline and contained adsorbed water. The synthetic and natural manganese oxide minerals with structural Mn^{4+} ions can act as oxidizing agents toward Fe^{2+} in solution (Postma, 1985). Under suitable pH's ($\approx$5), the Fe^{3+} will hydrolyze, polymerize and precipitate as iron oxides that are pseudomorphic after the initial manganese oxide mineral. This process can be exemplified by the following reactions:

$$Fe^{2+}(aq) \rightarrow Fe^{3+}(aq) + e^- \tag{1.15}$$

$$Fe^{3+}(aq) + 3H_2O \rightarrow FeOOH + H_2O + 3H^+ \tag{1.16}$$

$$MnO_2(s) + 4H^+ + 2e^- \rightarrow Mn^{2+} + 2H_2O \tag{1.17}$$

The cause for pseudomorphism appears to be the oxidation of Fe^{2+} at the site of the Mn^{4+} ion and Fe^{3+} ion being hydrolyzed and polymerized immediately following oxidation so that the immobile Fe-oxide phase efficiently replaces the prior position of the Mn-oxide. Ostensibly the rate of hydrolysis of Fe^{3+} (Eq. 1.16) controls the degree to which the pseudomorphic form of Mn-oxide is preserved. For good pseudomorphism the kinetics of hydrolysis of Fe^{3+} has to be fast otherwise the second phase would form independently due to movement of Fe^{3+} from the site of oxidation. The anions in the solution had some effect on the nature of the iron-oxide precipitated in that SO_4^{2-} resulted in feroxyhyte and then to goethite, perchlorate (with Cl- as an impurity) to lepidocrocite or akaganeite, and CI- to akaganeite. Formation of either ferrihydrite or feroxyhyte has been described in a stream receiving acid mine drainage rich in SO_4^{2-} ions (Brady et al., 1986). Comparable minerals were observed in ochreous deposits in drain pipes and ditches at neutral pH (Süsser and Schwertmann, 1983). Chukhrov et al. (1976) also described ready alteration of feroxyhyte to goethite. Because both feroxyhyte and ferrihydrite are intermediates instead of final products they are not anticipated to be observed widely in nature except in a Fe^{2+} solution that is currently undergoing oxidation. Little or no Mn substitution in the iron-oxide structure was indicated by the chemical analyses of the iron-oxides formed in the reactions. The presence of goethite together with

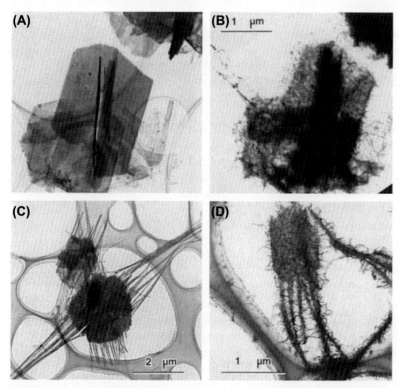

FIGURE 1.26 Transmission electron microscopy (TEM) of (A) birnessite, (B) iron oxide resulting from the reaction of the birnessite particle in (A) and $FeSO_4$, (C) todorokite formed from birnessite, (D) iron oxide resulting from the reaction of $FeSO_4$ and todorokite (Golden et al., 1988).

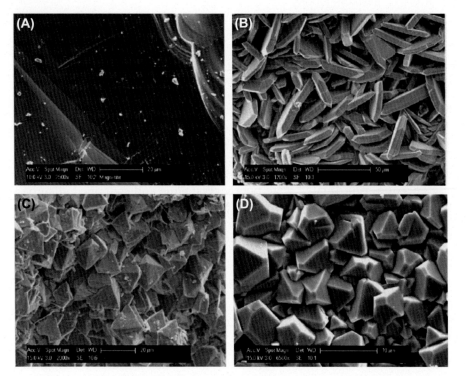

FIGURE 1.27   Secondary electron micrographs (SE) of (A) crushed magnetite crystals, (B) iron disulfides (predominantly marcasite), (C) and (D) pyrite formed on magnetite grain surface (Qian et al., 2010).

Mn-oxides might form via the intermediate ferroxyhyte or lepidocrocite. Formation of akaganeite was associated with the chloride anion and should not be expected in nature except in highly saline soils and in marine or comparable environments (Golden et al., 1988).

Qian et al. (2009, 2010) presented an experimental study on the sulfidation of magnetite to form pyrite/marcasite under hydrothermal conditions (90–300°C, vapor saturated pressures), a process related to gold deposition in several ore deposits. Marcasite was created only at $pH_{21°C}$ values below four and was the principal Fe disulfide at $pH_{21°C}$ 1.11, whereas pyrite dominated at $pH_{21°C}$ values above two and precipitated even under basic conditions (up to $pH_{21°C}$ 12–13) (Fig. 1.27). Marcasite crystallization was preferred at higher temperatures. Fine-grained pyrrhotite crystallized at the early stage of the reaction accompanied by pyrite in a few experiments with large surface area magnetite (grain size smaller than 125 μm). This pyrrhotite ultimately gave way to pyrite. The transformation rate of magnetite to Fe disulfide amplified with falling pH (at 120°C; $pH_{120°C}$ 0.96–4.42), and the rate of the transformation increased with temperature from 120 to 190°C. SEM imaging showed the existence of micropores (0.1–5 μm scale) at the reaction front between the initial magnetite and the product pyrite, and the pyrite and/or marcasite were euhedral at $pH_{21°C}$ values below four and anhedral at higher pH values. The newly crystallized pyrite was microporous (0.1–5 μm); this microporosity assists fluid transport to the reaction interface between magnetite and pyrite, consequently supporting the replacement reaction. The pyrite formed on the original magnetite was polycrystalline and did not maintain the crystallographic orientation of the magnetite. The pyrite precipitation was also found on the PTFE liner, which is in line with pyrite crystallizing from solution. The mechanism of the reaction is that of a dissolution–reprecipitation reaction with the crystallization of pyrite as the rate-limiting step relative to magnetite dissolution under mildly acidic conditions. The experimental results agree well with sulfide phase assemblages and textures described from sulfidized Banded Iron Formations: pyrite, marcasite and pyrrhotite have been observed to occur or coexist in various sulfidized Banded Iron Formations, and the microtextures show no proof of submicrometer scale pseudomorphism of magnetite by pyrite (Qian et al., 2010).

In experiments from 300 to 900°C and 200–1000 MPa monazite was both partly replaced and overgrown by $ThSiO_4$, BSE imaging, EBSD analysis, and TEM showed that in the experiments from 500 to 900°C, the $ThSiO_4$ phase appeared in the form of monoclinic huttonite suggesting that huttonite, accompanying monazite, might exist metastably over a much larger P-T range than formerly believed (Fig. 1.28). TEM analysis of a foil cut at right angles to the monazite–huttonite interface from the 600°C, 500 MPa experiment using a focused ion beam (FIB) suggests that the huttonite in addition to the border between the huttonite and monazite may be described by the presence of numerous fluid inclusions. High-resolution TEM analysis suggests that the huttonite–monazite interface is coherent. In the case of replacement of monazite by huttonite, fluid-aided dissolution–reprecipitation is anticipated as the most probable mechanism responsible. Huttonite has a lower P-T, tetragonal dimorph, thorite, which is substantially more common. At 1 bar, the experimental, unreversed phase transition between thorite and huttonite is situated between 1210 and 1225°C. At higher pressures, huttonite has been revealed experimentally to be stable over a range of temperatures usually encountered in the lower crust to upper mantle. In all experiments, a $ThSiO_4$

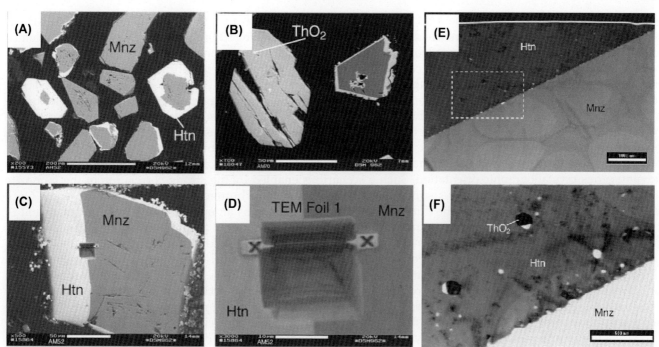

FIGURE 1.28 Back-scattered electron (BSE) photos of experimentally metasomatized monazite at 600°C showing examples of a monazite grains that have been nearly totally replaced by huttonite (A) and a monazite grain completely replaced by huttonite (B) with small inclusions of ThO$_2$. (C) BSE photos of experimentally metasomatized monazite with a huttonite rim (600°C, 500 MPa) showing the location of the FIB cut. The exact location of the foil is between the two spots marked with an X in (D). (E) Full extent of the FIB cut TEM foil (cf. C and D) is shown in (E) with a close up of the interface shown in (F). Large, elongate *circles* reflect holes in the carbon substrate upon which the foil rests. The dark band across the top of (E) is a portion of the 1 μm thick Pt layer. Note the sharp interface between the monazite and huttonite, the numerous fluid inclusions (now bright empty voids) in the huttonite, and dark inclusions, commonly associated with the fluid inclusions. The dark inclusions represent high concentrations of more electron-absorbing Th, in the form of ThO$_2$, which were included during the growth of the huttonite rim (Harlov et al., 2007).

phase (or phases) precipitated together with the monazite. In the 300°C experiments at 200 and 500 MPa, the ThSiO$_4$ phase occurred in the form of a fine-grained, polycrystalline mass consisting of randomly oriented crystals smaller than 1 μm partly surrounding a subset of the monazite grains besides growing along obvious cleavage planes or cracks in the monazite. In the 400–700°C experiments at 500 MPa and in the 900°C experiment at 1000 MPa, a ThSiO$_4$ phase incompletely replaced and/or overgrew the monazite around the grain rims. This replacement happened for 10%–20% of the monazite grains. BSE imaging and EMP traverses of these experiments suggest that the chemical margin between the two phases is compositionally sharp on the micrometer scale. In the 600°C, 500 MPa experiments, a couple of the smaller monazite grains were completely replaced by this ThSiO$_4$ phase as a pseudomorph while retaining the outward monoclinic symmetry of the original monazite grain. Incomplete replacement and/or overgrowth of monazite by a ThSiO$_4$ phase along the grain rim is comparatively more common in the 400, 500, 600, and 700°C experiments at 500 MPa and less common in the experiment at 900°C and 1000 MPa. In the 900°C experiments, such overgrowths are sporadic and, when they do occur, are more limited in scope. Instead, at 900°C and 1000 MPa, the majority of the Th and SiO$_2$ have reacted to form small (5–10 μm), isolated, euhedral, tetragonal grains of thorite. Overall, for each of the other experiments at lower temperatures, a percentage of the Th^{4+} and SO$_4^{2-}$ in solution formed as discrete crystals of thorite. In all cases, these crystals were confirmed to be thorite from their tetragonal symmetry (where obvious), EMP analysis, and EBSD patterns. In some metasomatized monazite grains (e.g., at 600°C and 500 MPa), the ThSiO$_4$ phase seems to have crystallized along preexperimental cracks (or cleavage planes?) in the monazite, such that isolated islets of monazite are left that are surrounded by the ThSiO$_4$ phase. In TEM the boundary between the huttonite and the monazite is quite sharp and is highlighted by numerous, small voids. These voids are thought to be fluid inclusions that were opened as a result of the FIB milling process. Although the monazite seems to be inclusion-free, the huttonite encloses both these fluid inclusions along with dark inclusions of ThO$_2$. In a lot of instances, the dark inclusions are linked with a partial void. Furthermore, the fluid inclusion density seems to demonstrate an ambiguous increase as the huttonite—monazite interface is approached. High-resolution TEM imaging suggests that at the boundary, the lattice fringes from the huttonite and monazite look to be continuous with little or no mismatch inferring that the interface between the huttonite and the monazite is coherent. This is proven by the electron diffraction pattern taken over the interface. There is no experimental proof that solid-state volume diffusion of Th and Si cations in monazite could make up the breadth and depth of the huttonite rims, the sharp compositional margin between the monazite and huttonite, the obvious replacement of monazite by huttonite in these rims, or even the replacement of whole monazite grains by huttonite. Creation of the huttonite rims must have happened by two potential mechanisms, both of which would elucidate the lack of a mismatch between the huttonite and monazite lattice fringes. Mechanism one comprises nucleation and growth of huttonite on those monazite crystal faces with the most advantageous crystallographic orientation. Such a scenario is enabled by the strong correlation of the monazite and huttonite lattice parameters, thus supporting epitactic growth of huttonite on a monazite substrate. Once the first atomic layer is laid down, the huttonite then grows outward as an overgrowth with no replacement of the monazite grain. The existence of any porosity in the huttonite rim could at that point be described as being due to the inclusion of fluid droplets by an outwardly moving huttonite—fluid interface. Mechanism two encompasses early crystallization of huttonite on the monazite crystal face, but at that point growing inward via dissolution of the monazite with replacement by huttonite. Such a process has to be, by definition, a coupled one because the rates of both the dissolution of the original mineral phase and the reprecipitation of the new mineral phase have to be identical so as to preserve contact between the reactants, products, and fluids, permitting for the transportation of material across the reaction front (Putnis, 2002). In such a process, the monazite is being replaced either incompletely or completely by a huttonite pseudomorph. The dissolution—reprecipitation process functions fundamentally as a fluid-aided chemical reaction, involving a Gibbs-free energy. The reactive volume is described by an ubiquitous fluid-filled porosity all through the metasomatized region and a sharp compositional margin between the freshly formed and initial mineral phase. This fluid-filled porosity permits rapid fluid-aided mass transfer to and from the reaction front at a rate about 10 orders of magnitude greater than simple volume diffusion through the crystal lattice (see discussion in Harlov et al., 2005). Sourcing the huttonite—monazite interface with Th and Si from the surrounding solution and transport of P and (Y + REE) from the interface into solution could have followed the general reaction:

$$(Ce, La, Nd, Pr) PO_4 + Th(NO_3)_{4(aq)} + H_4SiO_{4(aq)} = ThSiO_4 + (Ce, La, Nd, Pr)^{3+}_{(aq)} + PO^{3-}_{4(aq)} + 4HNO_{3(aq)}$$

$$\text{Monazite-(Ce)} = \text{Huttonite rim on Monazite} \qquad (1.18)$$

The destiny of any excess PO_4^{3-} and $(Y + REE)^{3+}$ in solution most probably plates out on unreacted monazite grains. Reaction (1.18) does not allow for the fact that a certain proportion of the Th^{4+} and SiO_4^{4-} in solution seem to have crystallized out as grains of thorite. Validation that dissolution—reprecipitation did occur, in combination with reaction (1.18), is observed both in the monazite grains partially replaced by huttonite, monazite grains almost or completely replaced by huttonite, in addition to the acidity of the solution at the conclusion of the experiment. Further corroboration of reaction (1.18) is also observed in the comparatively high solubility of Th in nitric acid. More significantly, both mechanisms one and two ensue in a P-T region, i.e., 500, 600, and 700°C at 500 MPa, and 900°C at 1000 MPa, where independent thermochemical data indicate that huttonite should be thermodynamically unstable compared to thorite. The perseverance of huttonite as a seemingly stable phase during the experiment (and in nature) might perhaps be due to the sluggish transformation of huttonite to thorite at such low temperatures. Whatever the case, the Gibbs free energy driving reaction (1.18) has to be negative over this P-T range. Mass transfer by diffusion via a fluid to and from the huttonite—monazite interface has to be achieved through a fluid-filled interconnected porosity within the newly crystallized huttonite. Proof for this porosity is seen in the large number of fluid inclusions in the huttonite, particularly at the monazite—huttonite interface. Analogous dissolution—reprecipitation related porosities or proof that such porosities was present before have been described in metasomatized FAP (Harlov et al., 2005), in feldspar (Putnis et al., 2007a), in tourmaline (Henry et al., 2002), in addition to a series of other metasomatized minerals (see review and discussion in Putnis, 2002). Probably once the dissolution—reprecipitation process stopped, such that the huttonite rim finished to grow, the rim recrystallized, thereby destroying the interconnected porosity and leaving isolated fluid inclusions. An additional option is that recrystallization of the huttonite rim and the following destruction of this porosity stopped the dissolution—reprecipitation process. In the current case, the difference in the molar volume between monazite (44.98 cm^3)

(for $CePO_4$) and huttonite (44.58 cm^3) is about 0.40 cm^3 or a reduction in volume of 0.88%. Such a decrease is possibly not sufficient to make up the total porosity required. Differences in the solubility of monazite and huttonite in the acidic solution must also have played a role. To enforce porosity on the crystallized huttonite phase, more moles of monazite must have been dissolved than moles of huttonite precipitated in such a way that reaction (1.18) was a molar deficit reaction (see disucssiondiscussion in Putnis, 2002). Such a reaction, together with a reduction in the molar volume from monazite to huttonite, could have been the mechanism behind the formation of an interconnected porosity in the huttonite rims. The results of this study could enlighten why huttonite seemingly comes from rocks that were metamorphosed at temperatures and pressures at which thorite would be anticipated to be the stable phase. $ThSiO_4$ huttonite grains were initially monazite grains with a possible nominal $ThSiO_4$-$CaTh(PO_4)_2$ component, which, at some point in their past, were metasomatized in a highly reactive, high-pH fluid disproportionately enriched in Th with stochiometrically equal amounts of accessible silica. The consequence of this metasomatism is that the ensuing huttonite grains are basically pseudomorphs of the original monazite grains. It also suggests that natural occurrences of huttonite mistaken for thorite are perhaps much more common than earlier recognized and that better attention must be paid to differentiating these huttonite grains from thorite using EBSD or Raman spectroscopy or a combination of both techniques (Harlov et al., 2007).

Neutralization of acidic fluids through fluid—olivine interactions is significant in volcanic environments and has been suggested as a useful scheme for the neutralization of acidic sulfate-rich fluids. To understand the interaction of olivine with highly acidic fluids King et al. (2011) studied the reaction of whole olivine crystals and a dunite cube with various sulfuric acid solutions at temperatures between 60 and 120°C. Reaction of olivine with 2 and 3.6 M acid concentrations created a layered amorphous silica pseudomorph of the original olivine grain (Fig. 1.29). The formation of a layered silica pseudomorph, the incorporation of ^{18}O into the silica rim and the dependency of the replacement rim thickness on the acid concentration indicate that the pseudomorphic replacement occurred through an

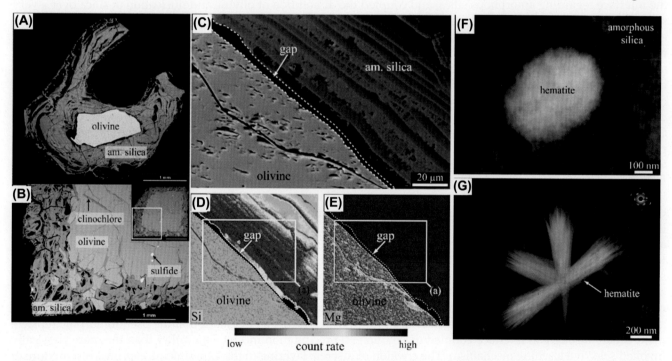

FIGURE 1.29 (A) Back-scattered electron (BSE) images of a cross section through a gem-quality olivine bead reacted with 3.6 M sulfuric acid at 90°C for 2 days. The grain has retained its original external morphology. The brighter contrast in the center is from the unreacted olivine core. (B) BSE image of part of the cross-section through the dunite cube (represented by the *white box* on the complete cross-section shown in the inset where the scale bar is 2 mm) reacted at 90°C in a 3.6 M sulfuric acid solution for 7 days. (C) BSE image of a silica—olivine interface, (D) Si, and (E) Mg compositional map of the olivine-amorphous silica interface, where the boxes in (D) and (E) outline the part of the cross-section shown in C. (F—G) Different types of hematite crystals found inside the amorphous silica phase. (F) High angle angular dark-field detector (HAADF) image of ellipse hematite particle and (G) HAADF image of agglomerated hematite particle with selected-area electron-diffraction ring pattern (inset). HAADF is sensitive to atomic weight. Therefore, in these images a brighter contrast represents a material containing elements with a large atomic weight hence higher Fe content in comparison to the surrounding silica (King et al., 2011).

interface-coupled dissolution—reprecipitation mechanism. A gap between the olivine precursor and the silica product is visible in the BSE image and as a black area in the element distribution map (Fig. 1.29C–E). It is possible in this case that the gap was formed due to shrinkage of the silica during drying, which pulled it away from the olivine surface. The differing colors on the silica map shows that the silica concentrations within the analyzed volume varied within the reaction rim. Silica layers near the bulk solution typically had lower silica concentrations than did a portion of the layer directly next to the gap and hence nearest to the olivine (visible as an orangey yellow area). Nevertheless, other areas of silica directly at the interface showed a high silica content. With exclusion of the layer directly at the interface, the innermost silica layers had the highest silica content. The relationship is intricate because the silica concentration varied in a nonsystematic way between the layers. As the silica is chemically pure, the generation of different SiKα intensities indicates that the discrete layers either contained different amounts of water or that the porosity of the layers fluctuates. Raman spectroscopy did not show substantial amounts of water in the silica product after drying, consequently the difference in the SiKα intensities can best be described by porosity variations. The observation that a pseudomorph was formed after experiments performed in 2 and 3.6 M sulfuric acid but not at lower acid concentrations indicates that there was a spatial coupling between olivine dissolution and silica precipitation in these experiments. In these experiments, the replacement of olivine by silica would follow reactions (1.19) and (1.20) for general olivine dissolution at acidic conditions where Eq. (1.20) exemplifies silica polymerization.

$$(Mg,Fe)_2SiO_4 + 4H^+ \rightarrow Mg^{2+} + Fe^{2+} + Si(OH)_4(aq) \tag{1.19}$$

$$n[-SiOH + HOSi-] \rightarrow SiO_2(s) + nH_2O \tag{1.20}$$

In this mechanism, the congruent dissolution of olivine is coupled in space and time to the precipitation of silica at an inwardly moving reaction front. In the experiments, the dissolution of olivine rapidly supersaturated the fluid at the interface regarding amorphous silica, and coupled the dissolution of olivine and nucleation of silica over a very short distance. This caused the formation of a pseudomorph of the original olivine grain. At the pH conditions of the experiments, the slow formation and destruction of siloxane bonds likewise precluded the growth of particles by means of Ostwald ripening. In its place, after the creation of discrete silica nanoparticles, aggregation would ensue as the nanoparticles collide with each other. This process was helped at low pH by the high concentrations of protons in solution that formed an active cationic complex on the nanoparticle surfaces. The aggregation of the nanoparticles created a network structure, which at that time altered into a gel. Preservation of individual grains in the dunite cube during the reaction with acid proves that the formation of the first, shape-defining gel layer has to be very fast. Consequently, gel solidification took place very close to the reacting olivine surface. This rapid solidification would preclude the released silica from merging with silica from other grains generating the distinct replaced grains. The initial olivine grain morphologies were retained in the dunite during replacement and formed voids in the core of the amorphous silica pseudomorphs, indicating that, notwithstanding the potential for amorphous silica to pacify the reaction, the olivine grain could be totally dissolved. This also infers that the interfacial fluid was replenished during the reaction through the very porous gel structure. In the experiments, a comparable change in the rate-limiting step was observed between the concentrated sulfuric acid experiments, hence very low pH, and lower acid concentration experiments. In the experiments in 2 and 3.6 M sulfuric acid, olivine dissolution was anticipated to be the rate-limiting step of the replacement process. Therefore, precipitation of the silica was closely coupled in space and time to corresponding olivine dissolution and a pseudomorph was formed. In contrast, in 1 M sulfuric acid, silica precipitation became the rate-limiting step and the dissolution and precipitation processes were no longer coupled over a spatial distance that could create a pseudomorph. The change in the rate-limiting step hangs on the pH and, hence, on the chemistry of the interfacial fluid, which is further evidence that the pseudomorphic replacement occurs through an interface-coupled dissolution—reprecipitation mechanism rather than the more classically invoked selective leaching mechanism. The creation of discrete layering in the experiments is probably a mixture of pre- and postnucleation effects. One possible model for the layering detected in the experiments could be the initial development of silica with varying densities, which were then separated to form distinct layers via shrinkage during silica maturation or as the coagulate formed during washing. Irrespective of the mechanism, the formation of layers suggests that the replacement layer is inhomogeneous. Additionally, it indicates that the silica layer formation has an influence on, but does not totally inhibit, further olivine dissolution in these systems. In the 0.1 M sulfuric acid experiments, the absence of alkali ions in solution would produce hydronium jarosite [$H_3O^+Fe_3(SO_4)_2(OH)_6$], as detected in hydrothermal experiments (Golden et al., 2008). In contrast, the experiments that contained Na^+ in solution would increase the precipitation of natrojarosite [$NaFe_3(SO_4)_2(OH)_6$] due to its lower solubility compared to hydronium jarosite. This is consistent with the experiments with 1 M sulfuric acid, where no Fe-phase was observed. However, when Na^+ was present in solution with the same original concentration of acid a yellow silica phase was formed. The yellow fluid formed from the reaction of olivine with partially neutralized sulfuric acid at 60°C was anticipated to result either from suspended precipitates of less than 0.45 μm or a Fe–sulfate complex in the solution, which usually forms a yellow solution. Jarosite precipitation is also boosted by increasing the temperature (Dutrizac, 1983). This effect can be seen in reactions with the pH = 2 solution, which formed hematite at 90°C but a yellow solution at 60°C. Changes in the final hematite color could be connected to the difference in preservation of sulfate within the precipitated hematite phase, which increases as temperature decreases. Irrespective of the preliminary phase, the formation of acidity during the hydrolysis of Fe ions in solution to produce the precursor complex (e.g., reaction 1.19 for jarosite precipitation; Dousma et al., 1979) would reduce the driving force for precipitation of an Fe-phase in extremely acidic solutions,

$$3Fe^{3+} + 2SO_4{}^{2-} + 6H_2O \rightarrow \left[Fe_3(SO_4)_2(OH)_6\right]^- + 6H^+ \tag{1.21}$$

This corresponds to the observed absence of Fe-phase precipitation in the experiments of olivine dissolution in 1 M sulfuric acid concentrations and above. When olivine was reacted with 1 M sulfuric acid amorphous silica was formed but no longer in the form of a pseudomorph of the olivine grain. Reaction with 0.1 M acid, or solutions containing Na, stimulated the creation of hematite along with amorphous silica. From the known Fe-phase stabilities for the experimental conditions in this study and the dependency of hematite formation on the presence of Na it was proposed that to begin with jarosite phases precipitated, which transformed into hematite during the experiment (King et al., 2011).

Mineral replacement reactions play a very important role in many natural processes and also have important technological applications. The pseudomorphic transformation of a material also has important technological applications, as it enables manufacturers to create a material with a predefined form and a controlled internal microstructure by choosing an appropriate initial material with the right shape and microstructural characteristics. An accurate control of the microstructural characteristics of a material (e.g., particle size, porosity) is critical because these characteristics determine its physicochemical properties and its performance during its end uses (Kocks et al., 2001). Alvarez-Lloret et al. (2010) for example, studied the a pseudomorphic mineral replacement mechanism involving a superficial dissolution of calcite and a subsequent overgrowth of oriented carbonated hydroxylapatite nanocrystals of sea urchin spine (*Paracentrotus lividus*) (Fig. 1.30). This replacement process was favored by an increase in porosity that enabled both fluid and mass to be transported by diffusion, thereby allowing the replacement reaction to proceed toward the inside of the spine. The epitaxial relationship observed between the parent calcite crystal and the newly formed apatite crystals was defined as (0001) apatite$//\left(01\bar{1}8\right)$ calcite and [10.0] apatite$//\left[\bar{4}4.1\right]$ calcite. The apatite crystals are related by the threefold axis arising from the trigonal symmetry of the original calcite crystal. There is, therefore, a strong structural control which favors the conversion of calcite into apatite.

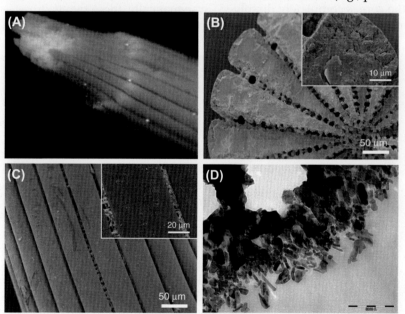

FIGURE 1.30 (A) Photograph of a transformed sea urchin spine showing the external shape and porosity of the original material. (B) Scanning electron microscopy (SEM) image of the cross section of a spine showing the well-preserved internal microstructure and its complex spongy morphology. (C) SEM image of the external morphology of the original calcitic sea urchin spine. (D) TEM image of converted sea urchin spines showing the newly formed apatite crystals, which are elongated and increase in size from the outer surface inward (reprinted with permission from Alvarez-Lloret, P., Rodriguez-Navarro, A. B., Falini, G., Fermani, S., and Ortega-Huertas, M., 2010, Crystallographic control of the hydrothermal conversion of calcitic sea urchin spine (Paracentrotus lividus) into apatite: Cryst. Growth Des., 10, 5227-5232. Copyright 2010 American Chemical Society).

GEOLOGICAL EXAMPLES

In this section a number of examples will be shown of pseudomorphism of a large variety of different minerals from a variety of different geological settings. They will be roughly divided in pseudomorphs in sedimentary settings, magmatic settings and metamorphic settings. It is in no way meant to be a complete overview as many more examples can be found in the literature, but it is the intention to show that this phenomenon is far more common than generally assumed by geologists and can occur at length scales from as small as a few micron up to several decimeters.

Pseudomorphs in Sedimentary Settings

Parent rock minerals are generally replaced by kaolinite and oxides of Al and Fe (and, occasionally, Mn) during lateritic weathering. The kaolinite is itself often also substituted by the oxides. The replacement is pseudomorphic and worldwide comprises huge volumes of parent rock. The three-dimensional relationship of textures forms through two coupled local reactions, a congruent dissolution above and a pseudomorphic replacement below. The dissolution of a grain is determined by H^+ supplied from above, and it releases aqueous. Al and/or Fe. Consecutively, this Al (or Fe) travels a short distance downward and forces the replacement of another grain by gibbsite (or hematite). Any silicate grain that develops pseudomorphic replacement by gibbsite or hematite involves import of Al or Fe, unless the porosity of the replacing aggregate is very high. Linking the condition for pseudomorphic replacement (i.e., the replacing mineral crystallizes at about the same rate as the replaced mineral dissolves) to reaction rate expressions leads to predicting "windows" in local water chemistry inside which pseudomorphic replacement is conceivable but outside of which it is not. During weathering, each mineral grain in the parent rock goes through a similar progressive order of alteration (inferred from the 3D sequence found in many profiles). First the grains are pseudomorphically replaced by oxide shells, oxide and parent mineral remain in contact (at least as viewed under the petrographic microscope), and shells of neighboring grains form "septa" with a central seam that indicates the original margin between the two parent grains. Then, at some time as the grain rises inside the profile, the replacement halts, and the grain core continues to dissolve creating a hollow space between itself and its pseudomorphic oxide shell. Finally, the core is completely dissolved. Systematically, these two textures—replacement followed upward by dissolution of grain cores—appear related in numerous profiles. For feldspars and pyroxenes this sequence usually happens over only a few millimeters from the fresh parent rock, and it can occur even within one grain length. The oxide shells form a strong framework mechanically able to carry the whole alteration profile without disintegrating, thus maintaining the overall bulk volume up into high levels of the profile (Fig. 1.31). This is why the saprolite forms most of the profile. The oxide framework, similarly resistant chemically, ultimately dissolves out in the nonsaprolite. After rising through a certain depth interval, which might or might not be substantial and inside which the oxide boxwork remains hollow, and as the voids approach the nonsaprolite, they become partially filled up with a second generation of oxides (Gibbsite II and Hematite II cements). Parent mineral grains can also be replaced by kaolinite (or other phyllosilicates), in which circumstance the kaolinite itself, higher in the profile, converts to oxides, showing the same order of partial replacement followed upward by nonreplacive dissolution that leaves voids. Contrary to pyroxenes and feldspars, however, this textural sequence takes place, for kaolinite, over a distance

FIGURE 1.31   Alteration or labradorite (Pl, gray): at center and bottom right a large labradorite grain is partly replaced by septa of gibbsite (white) with center seams; fragments are in optical continuity. At center, unreplaced labradorite has started to dissolve congruently, leaving voids (black). Toward top left labradorite is all dissolved away, leaving large voids. Replacement passes to dissolution over a grain's length. Sample from the Mt. Tonkoui Norite (labradorite, two pyroxenes. magnetite), Ivory Coast (Delvigne, 1965; photo 5, p. 160). Height of photo 0.5 mm (Merino et al., 1993).

of tens of meters. Up to the point in the profile where the replacement stops, the replacement of, for instance, plagioclase by gibbsite unavoidably needs the introduction of Al for it to be pseudomorphic, though the gibbsite aggregate includes significant microporosity. Any silicate that is pseudomorphically replaced by any one-cation oxide or hydroxide, such as gibbsite and hematite, entails import of that cation (unless the microporosity of the replacing aggregate is very high).

$$CaAl_2Si_2O_8(An) + Al^{3+} + 5H_2O = 3Al(OH)_3(Gibb) + Ca^{2+} + 2SiO_2(aq) + H^+. \tag{1.22}$$

$$An + (1 - 3\phi)Al^{3+} + \cdots = 3(1 - \phi)Gibb + \cdots \tag{1.23}$$

$$An + 8H^+ = 2Al^{3+} + Ca^{2+} + 2SiO_2 (aq) + 4H_2O \tag{1.24}$$

It seems clear that the Al^{3+} released by a dissolving a plagioclase core through reaction (1.24) travels a short distance downward and drives the replacement (reactions 1.22 or 1.23) of another plagioclase grain. The two reactions (1.22)–(1.24) are coupled through the aqueous Al. This is the reason why the two textures (dissolution voids just overlying replacements) are themselves inexorably linked in space (Merino et al., 1993).

Anand and Gilkes (1984) studied the chemical behavior of a variety of elements in a lateric saprolite formed from granite by lateritic isovolumetric chemical weathering. It contained pseudomorphs of various primary mineral grains. Most of the Si, Ba, Sr, Pb, and Cu present in feldspars in granite was lost on weathering to halloysite, whereas Ga, Al, and V were retained. Most of the Zn, Mn and Co that was present in biotite in granite was also lost during alteration, whereas the pseudomorphs retained Fe, Cr and Ni. Zn, Fe, Mn, Cr, and V present in magnetite were wholly retained within the hematite pseudomorphs, whereas Ni was lost. The plagioclase and alkali feldspars altered to a cloudy material that was subsequently replaced by a fine-grained porous pseudomorphous mixture of halloysite and lesser gibbsite. Plagioclase feldspars were observed to alter more rapidly than the alkali feldspars. Massive biotite grains have altered to exfoliated flakes composed mostly of parallel-oriented kaolinite crystals with the opened cleavages being partially filled with goethite. Massive magnetite grains have altered to an oriented arrangement of platy hematite crystals. In each case the alteration products occupy the space formerly occupied by the parent mineral grain, and as a consequence of the loss of much of their mass the alteration product is highly porous. This high degree of porosity is an inevitable consequence of isovolumetric weathering where introduction of elements via solution or suspension is minor or nonexistent.

Early diagenetic K-feldspar in the Triassic Buntsandstein of the Iberian Range (Spain) occurs as pseudomorphs after detrital K-feldspar (Or < 93) and plagioclase (Ab < 96). These pseudomorphs are chemically pure (Or > 99), without twinning, usually heavily clouded by vacuoles and miniscule inclusions, dark greyish blue-luminescence and are composed of numerous fine euhedral crystals of K-feldspar (Fig. 1.32). The latter property indicates that the pseudomorphs formed via dissolution of detrital K-feldspar and plagioclase and precipitation of authigenic K-feldspar. The variation in chemical composition between the diagenetic and the K-feldspars ($Or_{99.7}Ab_{0.3}An_{0.0}$ and $Or_{80.9}Ab_{17.8}An_{1.3}$, respectively) was measured by means of BSE imaging and X-ray elemental distribution of K and Na. Authigenic K-feldspar was continuously virtually pure end member (Or > 99 mol%), whereas the unaltered K-feldspars contain Na and Ca impurities. Under the scanning electron microscopy (SEM), K-feldspar pseudomorphs contain ample dissolution pits and discrete pseudorhombic and prismatic crystals (1–60 μm long) that are aligned parallel to one another, and have even surfaces (Fig. 1.32). Unaltered detrital K- feldspar and plagioclase

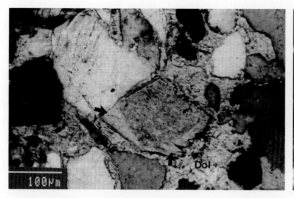 

FIGURE 1.32   (Left) K-feldspar pseudomorph that is untwinned, clouded by tiny inclusions and vacuoles and has a clear K-feldspar overgrowth (*arrows*); cross-nicols. (Right) A scanning electron microscopy (SEM) micrograph of a K-feldspar pseudomorph composed of tiny euhedral crystals of diagenetic K-feldspar (e.g., *arrows*); RKF is relict detrital K-feldspar (Morad et al., 1989).

exhibit etched surfaces due to diagenetic dissolution. In various cases, the pseudomorphs consist of many elongate, rod-like prismatic crystals. Several lines of evidence indicate that the K-feldspar pseudomorphs of the Buntsandstein were formed diagenetically. These are: (1) Authigenesis of K-feldspar was limited to the transitional and marine facies of the Buntsandstein. This feature indicates that authigenesis of K-feldspar was favored under a definite set of geochemical conditions (primarily a high a_{K^+}/a_{H^+} ratio and a high $a_{H_4SiO_4}$). (2) The less common incidence of diagenetic replacement by K-feldspars in early diagenetic concretionary cemented parts of the sandstones. Under such conditions the detrital feldspars were inaccessible from the geochemical system which favored crystallization of K-feldspar. (3) The practically pure end-member composition likewise point to a diagenetic origin for the K-feldspar pseudomorphs. Detrital K-feldspars of the Buntsandstein contain substantial amounts of, predominantly, albite solid solution. (4) The dark cathodoluminescence of the K-feldspar pseudomorphs is characteristic of a diagenetic origin as well. (5) The presence of microporosity (vacuoles) and inclusions in the K-feldspar pseudomorphs are typical of diagenetic feldspar pseudomorphs. (6) The euhedral shapes and flat surfaces of the K-feldspar crystals in the pseudomorphs and overgrowths are similar to diagenetic albite pseudomorphs and overgrowths, therefore their diagenetic origin. (7) Obvious lack of twinning is distinctive for diagenetic feldspar pseudomorphs. This character is thought to be related to the fact that the textural framework of feldspar pseudomorphs generally consist of abundant discrete smaller crystals rather than a single crystal. Instead of twinning, some of the pseudomorphs are described to have a patchy extinction pattern that is caused by variations in the optical orientation of authigenic K-feldspar domains or the presence of relict detrital feldspar. The existence of diagenetic K-feldspar as abundant usually parallel aligned crystals with variable amounts of intercrystalline porosity suggests that pseudomorphism takes place via a partial dissolution−precipitation process which promotes nucleation on numerous sites. The first step involves instantaneous exchange of alkali ions (in this case H^+ exchanged for K^+ and Na^+) followed by formation of Al-Si chain structures before diffusion of Si and Al as soluble ions occurs. The replacive authigenic K-feldspar would therefore have subsequently crystallized by nucleation on these Si-Al structural units. The intercrystalline porosity in the pseudomorphs has seemingly acted as conduits for outwards diffusion of Na^+ and Ca^{2+} and inwards diffusion of K^+ to the detrital feldspar-pore solution interface. This has boosted the penetrative replacement by K-feldspar that in the end resulted in a complete pseudomorph. The control of the replacement progression by grain fractures as well as twin and cleavage traces is firstly because these surfaces act as weakness planes along which pore fluids can penetrate the detrital feldspar, and secondly, is possibly because these surfaces are considered to have larger free energy, thus higher dissolution rates. The diagenetic geochemical conditions that favored crystallization of authigenic K-feldspar is apparently connected to meteoric waters recharged in the permeable alluvial conglomeratic sediments at the base of fault escarpment. These waters evolved chemically (a_{K^+}/a_{H^+} and $a_{H_4SiO_4}$ increased) through reactions (nearly complete leaching or kaolinitization) with detrital K-feldspar and micas. Authigenesis of K-feldspar has occurred at atmospheric T-P conditions (Morad et al., 1989).

Blank and Fosberg (1991) investigated secondary minerals in a duripanson the Owyhee Plateau of south-western Idaho. Within this soil horizon, SiO_2 was released through the alteration of primary aluminosilicate minerals and migrated into the duripan laminar cap via the addition of percolating Si-rich solutions from the overlying soil horizon. Initially the SiO_2 occurred as gel-like materials that later reorganized into a near closed-packed arrangement of SiO_2 sol spheres or opal-A. Calcite in the form of scalenohedra occurred within some isolated units. In these instances, the morphological outer shape of opaline silica structural units was similar to the calcite scalenohedrons, and the undulose extinction pattern of the opaline silica was in optical continuity with the extinction pattern of the calcite. Clear-opal (opal-A), therefore, appears to represent a pseudomorphic replacement of calcite pendants, whereas pore fillings dirty-opal represents alteration of primary minerals by the conditions within the duripan.

Glendonites, calcite pseudomorphs after the metastable mineral ikaite ($CaCO_3 \cdot 6H_2O$), occur in the Late Aptian interval of the Bulldog Shale in the Eromanga Basin, Australia and in other Early Cretaceous basins at high paleolatitudes (Fig. 1.33). Ikaite precipitation in marine environments necessitates near-freezing temperatures (not higher than 4°C), high alkalinity, increased levels of orthophosphate, and high P_{CO_2}. Ikaite has been detected naturally only at temperatures between −1.9 and 7°C in both marine and continental waters that are highly alkaline and rich in orthophosphate (Fig. 1.34). In a marine environment ikaite grows displacively at or just below the sediment−water interface in organic-carbon-rich sediments. Given time, ikaite will be replaced by calcite even at near-freezing temperatures. However, the transition to calcite has been observed to occur when temperatures rise above a certain range; the exact temperature depending on the water chemistry. Furthermore, it has been observed that the temperature at transition (or more precisely the magnitude and rate of the temperature increase) is a controlling factor in the preservation of the ikaite morphology. These observations indicate that if transition occurred in response to temperature change, ancient pseudomorphs must have experienced gradual temperature increases in order to have preserved the ikaite morphology. For aqueous environments conducive to ikaite precipitation, one would expect

FIGURE 1.33 A glendonite, calcite pseudomorph after ikaite, from the late Aptian Bulldog Shale at Petermorra Creek, South Australia. 1 cm (De Lurio and Frakes, 1999).

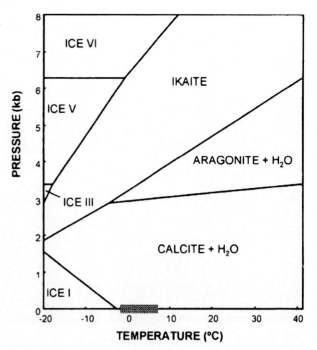

FIGURE 1.34 Generalized pressure—temperature phase relations for the calcium carbonate system in fresh water illustrating the stability range of ikaite. In a sea water system (lower activity of water), the stability fields are shifted to the right. At surface pressures, ikaite is optimum at low temperatures. Natural ikaite has been observed metastably at temperatures between −1.9 and 7°C (*shaded region*). To achieve precipitation at surface pressures at any temperature requires both elevated alkalinity to achieve supersaturation and elevated phosphate to inhibit calcite/aragonite precipitation. *Modified from Marland (1975) by Bischoff et al. (1993a) in De Lurio, J.L., Frakes, L.A., 1999. Glendonites as a paleoenvironmental tool: implications for early cretaceous high latitude climates in Australia. Geochim. Cosmochim. Acta 63, 1039—1048.*

transition near 8°C or lower as no natural ikaite has ever been observed to persist above this temperature and marine ikaite has never been reported at temperatures above 4°C. Larsen (1994) proposed that the ikaite to calcite transition, represented by the reaction

$$CaCO_3 \cdot 6H_2O \rightarrow CaCO_3 + 6H_2O \qquad (1.25)$$

results in a 68.6% volume loss as the six structural water molecules are released from the ikaite crystal. He identifies zoned calcite grains similar to those identified in the Bulldog Shale glendonites under Cathode Luminescence. In his model, the zoned grains represented initial transition calcite and occupy 31.4% of the pseudomorph. He attributed the remaining volume of calcite to later infilling. When calcite forms from an ikaite precursor there are two possible water sources with which the resultant calcite could reequilibrate: structural water and pore water. Laboratory experiments in which ikaite was replaced by calcite, whether surrounded by water or exposed to dry air, showed that replacement can be achieved by the ikaite structural water alone (Johnston, 1995 referred to as an "auto-pseudomorphing" fluid; Shaikh and Shearman, 1986). Trace element concentrations within a closed system vary spatially with time as the formation of the crystal proceeds and water chemistry changes; conversely, a crystal grown in an open system shows uniform composition (Frank et al., 1982; Marshall, 1988). The diminishment of the concentric zoning into more uniform orange luminescence implies that the ikaite structure had sufficiently collapsed to allow infiltration by pore water (De Lurio and Frakes, 1999).

Calcite pseudomorphs have replaced euhedral ikaite ($CaCO_3 \cdot 6H_2O$) porphyroblasts in Dalradian calcareous slates and metadolostones of western Scotland, with a volume decrease of at least 47% (Dempster and Jess, 2015). Porphyroblast—fabric relationships indicate that the initial growth of ikaite is older than a penetrative tectonic fabric formed during upright folding. This is the first reported occurrence of metamorphic ikaite porphyroblasts and points toward growth within the slates during an ultralow-temperature metamorphism with an exceptionally low geothermal gradient. This event is associated with the penetration of long-lived and extreme permafrost deep into subaerially exposed bedrock during Neoproterozoic glaciation. The porphyroblasts are pseudomorphs with a concentric structure typically with a c. 100—500 μm outer fringe of quartz and a center of calcite ± quartz ± pyrite or quartz ± pyrite (Fig. 1.35). The outer rim of inclusion-free quartz tends to be wider (up to 500 μm) at the

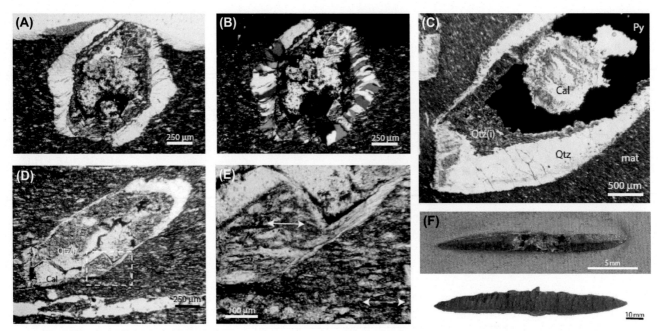

FIGURE 1.35   Thin-section photomicrographs of ikaite pseudomorphs; plane-polarized light unless stated otherwise. (A) Clear quartz rim around inclusion-rich calcite core, with minor replacement of calcite by inclusion-rich quartz. (B) Cross-polarized light view of (A) showing quartz fibers and rotational symmetry. (C) Ikaite pseudomorph with calcite (Cal) core, pyrite (Py) mantle and quartz (Qtz) rim; clear quartz rim surrounds inclusion-rich quartz (Qtz(i)). (D) Pseudomorph with clear quartz rim and calcite and inclusion-rich quartz core with concentration of Fe-oxides on grain boundaries between Cal and Qtz. *Box* shows location of (E). (E) High-magnification image of pseudomorph edge showing similar alignment of inclusions within quartz (*central arrow*) and alignment within matrix (*lower right arrow*). (F top) Large elongate ikaite pseudomorphs with pyramidal terminations from the Easdale Slates. (F bottom) Specimen of ikaite pseudomorph dredged from River Clyde near Cardross (calcite after ikaite, courtesy of Glasgow Museums Collections, G1988-50) (Dempster and Jess, 2015).

pyramidal terminations. Such quartz is commonly fibrous, and the fibers may be intergrown with scant chlorite and exhibit a curved morphology with a rotational symmetry. Other rims of clear quartz are coarser grained, especially those around cores dominated by either inclusion-rich quartz or pyrite. The pseudomorphs have sharp contacts with the matrix and typically there is no apparent disruption of the fabric in the matrix near the porphyroblasts. Calcite in the center of the pseudomorphs is granular and commonly contains graphite and Fe-oxide inclusions and more rarely chlorite, muscovite, and tourmaline. No relics of original porphyroblast minerals are preserved, although they are likely to be chemically similar to calcite, the first phase of replacement. The quartz fringes that surround the pseudomorphs have been interpreted as a growth from a Si-saturated fluid that filled the cavities formed during a volume loss associated with the initial replacement reaction. On the basis of 18 measured pseudomorphs, the solid volume loss has been estimated at $47 \pm 9\%$. This is likely to be a minimum as the pure transformation of ikaite to calcite would result in a maximum volume decrease of 68.6% (Larsen, 1994). The pseudomorphs' distinctive elongate shape with scalenohedral terminations, the initial replacement by calcite and their presence in organic-rich calcareous slates all point toward the porphyroblast being glendonite, a pseudomorph of hydrated calcium carbonate, ikaite. Thus both original growth of the ikaite porphyroblasts and their subsequent alteration or replacement are metamorphic features linked to ductile deformation. Ikaite pseudomorphs have not previously been reported as a metamorphic porphyroblast. Synsedimentary ikaite pseudomorphs in the geological record are typically well preserved, but examples within metamorphic rocks, such as those reported from Dalradian rocks in Ireland by Johnston (1995) rarely survive metamorphic events. In contrast, the pseudomorphs from the Easdale slates are well preserved, owing to the porphyroblast growth occurring within lithified metamorphic rock and the development of a protective quartz rim during the initial volume change. The growth of large euhedral ikaite porphyroblasts within the bedrock slate would almost certainly be favored by the presence of liquid water, required to counter the kinetic sluggishness as a result of low temperatures. This, together with evidence for growth postdating ductile deformation, seems to indicate that growth in the slate most likely occurred at depth in the crust. This is a unique occurrence of ikaite and suggests that an unusual set of circumstances was responsible for its growth. An extreme lowering of the surface temperatures after initial ductile deformation and during orogenesis may explain the presence of ikaite. Permafrost conditions are known to extend into the crust to depths of up to 1 km during glacial events lasting around 1 Ma, with thermal disturbance to several kilometers depth associated with Holocene glaciation (Šafanda et al., 2004). However, a "normal" glaciation is unable of influencing temperatures in the metamorphic realm. Only the combination of extreme very cold surface air temperatures, long time scales, and an absence of thermally blanketing ice cover is capable of lowering metamorphic temperatures at depth. The low temperatures may also favor more rapid rates of heat transfer in the shallow crust and hence a lower geothermal gradient (Dempster and Jess, 2015; Dempster and Persano, 2006).

The Alpine Haselgebirge Formation characterizes an Upper Permian to Lower Triassic evaporitic rift succession of the Northern Calcareous Alps (Eastern Alps). Although the rocksalt body deposits are highly tectonically affected, containing primarily protocataclasites and mylonites of halite and mudrock, the early diagenetic history can be recognized in nontectonized mudrock bodies: Centimeter-sized euhedral halite hopper crystals formed as displacive cubes within mud just in the course of shallow burial. The crystals were deformed by the following compaction. Later, migrating fluids resulted in the replacement of halite by anhydrite keeping the shapes of deformed halite cubes. Halite hopper crystals were taken from blocks of mudrock by Leitner et al. (2013). In characteristic samples, halite cubes measured from 2—4 mm to 10—30 mm in size. The biggest salt cubes in Berchtesgaden and in Altaussee were around 60 mm (edge length). The cubes usually occur within clusters of similar size in blurred layers, whereby the single halite hopper crystals are not in contact with one another. The external shapes of the crystals are cuboids, rhombohedrons, or rhombic prisms. The halite hopper crystals show concave crystal surfaces with distinct edges (hopper shape), whereby the surfaces are regularly stepped on a small scale. The larger the diameter of the "cubes", the more elongated their edges are. A red coating covering the cubes causes the typical red color, but within the halite crystals are clear. Some of the hopper crystals, particularly the bigger ones, enclose fragments of mudrock. Anhydrite cubes typically occur in fragments of internally undeformed mudrock. The diameter of the cubes varies between 7 and 200 mm (typically 30—40 mm) and the cubes often show elongated edges (hopper shape). The shear strength of consolidated mudrock transferred the stress required to deform halite. The glide and climb mechanism of halite operates at less than 1 MPa differential stress at low temperatures. The presence of water accelerates the intracrystalline deformation. Compaction of the mud and deformation of the halite crystals was complemented by the removal of excess pore water from the mud. The halite hopper crystals were in all probability deformed in a simple overburden-dominated stress field. The subsequent shape depends on the initial random orientation of the halite cubes, and cuboids, rhombohedrons and prisms develop. X-ray computed tomography (CT) observations showed a major finite strain orientation and two equal minimum and intermediate finite strain orientations. This axial shortening points to a compaction, where the highest strain orientation was parallel to gravity. Finally, rock stiffening by mechanical and chemical compaction preserved the halite cubes from extra compression and destruction. Anhydrite cubes were inferred as pseudomorphs after halite. Transformation clearly ensued after the halite cubes had reached their final size and shape. Calcium sulphate-rich solutions migrated into the cubes and replaced former halite (Fig. 1.36). Reasons for this explanation are as follows: (1) anhydrite crystals do not form cubes; (2) anhydrite crystals with elongated edges like halite are not known; (3) transition phases of halite cubes incompletely replaced by fibrous anhydrite growing inward toward the center were observed; and (4) electron microprobe analysis showed marginally higher concentrations of Na in anhydrite cubes suggesting a halite precursor. The pore water, which dissolved halite, was clearly oversaturated with respect to anhydrite. This dilution effect reveals dewatering of mud (rock). Brines undersaturated in connection with halite moved through the rock when mudrocks and anhydrite rocks were intact and not yet completely destroyed by the formation of the Haselgebirge mudrock—halite tectonite.

The Late Permian Salado Formation encompasses an elaborate range of both primary (including syndepositional diagenetic) and secondary minerals, textures, and fabrics. Primary textures and fabrics were documented for saline minerals such as gypsum

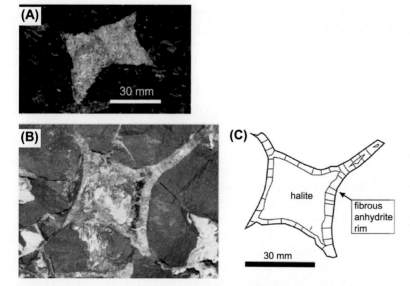

FIGURE 1.36 (A) Anhydrite cubes in undeformed mudrock of "bedded nodular" anhydrite structure. (B and C) Halite cube with strongly elongated edges surrounded by an anhydrite rim (Leitner et al., 2013).

(now pseudomorph replaced by anhydrite, polyhalite, halite, and sylvite as vertically oriented prisms, wave-rippled grains, mud-incorporated euhedra, and laminae made of small crystals) and halite. Pseudomorphs after primary gypsum were observed in laminated anhydrite and polyhalite, whereas primary halite was found in mud-free halite (both banded and massive) and in muddy halite. In one core sample, glauberite pseudomorphs within laminated anhydrite were changed to halite and anhydrite. These pseudomorphs have been tentatively interpreted as of early diagenetic origin. Laminated anhydrite was found in the McNutt Zone of the Salado Formation in 0.3—6 m thick beds. Laminae, characterized by repetitions of waxy gray anhydrite and dull gray magnesite-rich mud and silt, are flat parallel-sided, wavy, crinkly, and distorted. Inside and cutting through laminae are gypsum crystal outlines of variable morphology and fabric, now made up of anhydrite, polyhalite, halite, and sylvite, all pseudomorphous after gypsum. Primary sedimentary structures with gypsum crystal pseudomorphs were detected in thin sections and slabs. Euhedral and subhedral crystals with outlines of gypsum morphologies are now anhydrite, polyhalite, halite, and sylvite. They are prismatic, from 1 to 15 mm in length, and subvertically oriented, some with swallow-tail twinning and all coming from a common substrate. In further examples, vertically oriented gypsum pseudomorphs (now anhydrite) included mud that is ordered parallel to crystal faces as bands, seemingly representing growth stages of gypsum. Mud laminae completely included within gypsum pseudomorphs can occasionally be followed continuously through some crystals. Mechanically reworked gypsum pseudomorphs were found in thin beds containing anhydrite, polyhalite, halite, and sylvite, all pseudomorphous after well-sorted, internally cross-laminated euhedral and subhedral crystals of prismatic gypsum, 1—3 mm in length. These prismatic crystals were observed with their long axes parallel to the strike of cross-bed foresets or aligned with long axes down the foreset dip direction. Randomly oriented incorporative gypsum pseudomorphs were observed as euhedral and subhedral equant hemipyramidal crystals, now anhydrite, but interpreted as having been gypsum originally, outlined by gray magnesite-rich mud and silt matrix. These pseudomorphs after isolated gypsum crystals, 1—5 mm in cross-section, are randomly oriented, often with muddy centers and clear rims. Anywhere these gypsum crystal pseudomorphs are profuse, the lamination of the host sediment is partially to completely disturbed. Laminated gypsum pseudomorphs were found as laminae, from less than 1—5 mm in thickness, which comprise randomly oriented or aligned gypsum crystals in the form of subhedral prisms or subhedral to anhedral equant crystals, pseudomorphically replaced by anhydrite and polyhalite. Magnesite-rich mud is commonly interstitial to the pseudomorphed gypsum crystals, as well as in discrete laminae. Randomly oriented euhedral crystals, up to 2 cm in cross-section, with glauberite morphology, but now consisting of anhydrite and halite, was found in a 10 cm interval in one core, out of a total of 12 cores surveyed (Fig. 1.37). These glauberite pseudomorphs are randomly oriented and include anhydrite-rich and magnesite-rich mud laminae without noticeable displacement of the host sediment. Zoning is defined by variable amounts of sediment included within the crystals. Complete sequences range from 1 to 11 m in thickness. In its least mineralogical altered state, a sequence from bottom to top consists of: (1) basal gray-green mud, grading into; (2) thinly laminated anhydrite and magnesite, grading into; (3) thinly laminated anhydrite and magnesite with vertically oriented gypsum pseudomorphs, mud incorporative gypsum pseudomorphs, wave ripples composed of gypsum pseudomorphs, and gypsum crystal pseudomorph laminae; (4) banded halite with alternations of polyhalite laminae and halite beds; (5) massive halite with disseminated polyhalite; (6) muddy halite with mud most abundant in upper portions (Lowenstein, 1982).

Widespread quartz pseudomorphs after evaporitic minerals are interbedded with stromatolites in 2.2 Ga old sedimentary rocks in the Yerrida rift basin of Western Australia. These deposits preserve various original crystal

FIGURE 1.37 Randomly oriented pseudomorphs after zoned glauberite anhedra, now composed of microcrystalline anhydrite (gray) and halite (black) which define zoning in pseudomorphs (viewed under cross-nicols). Scale bar is 5 mm long (Lowenstein, 1982).

morphologies that grew displacively either as individuals or as clusters within stromatolitic horizons and accompanying fine-grained siliciclastic beds. Although the evaporitic minerals have largely been replaced by quartz, their crystal shapes including lenticular, rosette, needle, and nodular forms, suggest that they formed from calcium sulfate-rich brines. Original calcium sulfate minerals such as gypsum or anhydrite are indicated by the morphologies of the pseudomorphs; which is supported by the observation of anhydrite inclusions in the quartz crystals that form the pseudomorphs (Fig. 1.38). Crystals of original evaporitic minerals are commonly replaced by mosaics of quartz crystals of much coarser size than those of the matrix. The larger pseudomorphs display swallowtail twinning texture. Individual crystals of quartz, particularly within euhedral pseudomorphs, reach 3 mm in size. In the pseudomorphs, minute relict calcium sulfate in the form of crystalline anhydrite can be readily recognized within the quartz crystals and in the stromatolitic matrix. Preservation of original crystal shapes of evaporitic minerals suggests that replacement took place early in the diagenetic history of the sequence. Silicification of the original evaporites probably began early in the burial process, either in the zone of active phreatic flow or in the upper part of the zone of compactional flow, on condition that the matrix permeability was retained. The silica required for the formation of quartz and chert was most likely provided by synsedimentary hydrothermal or hot-spring input into the rift basin area. Chert with spherulitic and mosaic structures suggests primary chemical precipitation (in situ) of silica at the site of deposition. The original morphologies and delicate textures of these evaporites are well preserved and display crystal growth positions. Such replacement suggests that (1) silica was available in the early diagenetic environment, (2) replacement occurred under synsedimentary to early burial conditions, and (3) basin tectonics and thin sedimentary cover in the basin area allowed movement of hydrothermal fluids into the depositional area. Silica precipitated mainly from evaporative concentration and cooling of highly alkaline spring water, causing early replacement of microbial mats and microbial substrates (El Tabakh et al., 1999).

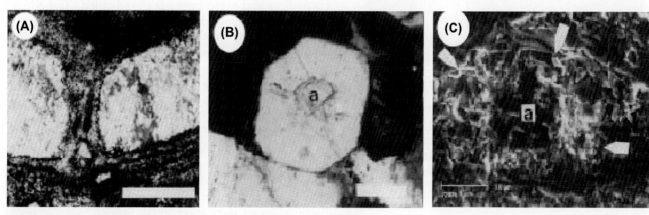

FIGURE 1.38 Petrography of evaporite replacement textures. (A) Twin of lenticular evaporite pseudomorphs (swallowtail crystals) of original gypsum grew displacively within stromatolitic layer and was replaced by microcrystalline quartz. Scale bar is 1 mm. (B) Photomicrograph showing *rectangular shaped* anhydrite crystal (a) within *hexagonal-shaped* quartz crystal. Scale bar is 0.25 mm. (C) Back-scattered electron image showing euhedral and blocky anhydrite (a) exhibiting typical 90 degrees cleavage. *White arrows* mark outline of crystal (El Tabakh et al., 1999).

Palaeoproterozoic chert in the Bartle Member of the Killara Formation, Yerrida Group, Yerrida Basin of Western Australia, contains an atypical association of crystal structures and rock fabrics that appears to indicate an evaporitic-pyroclastic-thermal-spring environment associated with rifting at about 2.2 Gyr. The chert member comprises silica pseudomorphs after evaporite minerals that in places surround relict isolated crystals and aggregates of crystals of gypsum and anhydrite. The evaporite minerals are accompanying minerals such as barite and analcime. Planar-laminated chert usually has nodular concretions up to 5 mm in diameter. The nodules are silicified and also have septarian cracks filled with chalcedonic quartz. These might be pseudomorphs after anhydrite formed through dehydration of gypsum. The rim of the nodules is characterized by prismatic crystals or bladed crystals; maybe pseudomorphs after anhydrite. Relict inclusions show the initial minerals were gypsum or anhydrite. Palmate, bladed, spindle-shaped crystals, and rosette-like or stellate aggregates are either silicified or replaced by kaolinite or Fe oxides (Fig. 1.39). Direct precipitation of silica from seawater was possibly more common in the Precambrian than in current environments. It is generally associated with carbonates and evaporites, and is probably to have happened in tidal flats and lagoons, where high evaporation rates produced high silica concentrations. The joint

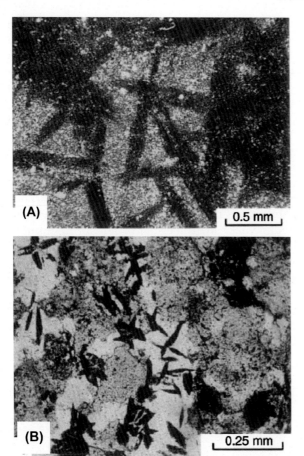

FIGURE 1.39 Plane-polarized light photomicrographs showing (A) kaolinite pseudomorphs of bunches of acicular crystals (possibly gypsum, or anhydrite) in a completely silicified matrix (microcrystalline quartz); (B) Fe oxides pseudomorphs of spindle-shaped crystals, probably gypsum, in a microcrystalline quartz matrix (Pirajno and Grey, 2002).

existence of organic matter and silica spheroids in the Bartle Member indicates that biogenic mediation can have played a role in the precipitation of some of the Bartle Member chert. In volcanic settings, thermal springs having dissolved colloidal silica produce prevalent silicification of bacterial mats. They precipitate laminae or sheets of amorphous silica that later recrystallize as chalcedony or crypto- and microcrystalline quartz. The conservation of organic material and putative microfossils in the Bartle Member indicate initial replacement by fine-grained silica introduced by mineralizing fluids. Especially, an outcrop of the Bartle Member about 90 km west of the main outcrops has comparable replacements of evaporite crystals and laminated stromatolitic chert, that closely look like the structures recorded from the Rietgat chert. The Bartle Member chert has many parallels with the Rietgat cherts, i.e., stratigraphic position at the top of a mafic volcanic succession, the structure of the chert (massive to laminated), and the occurrence of spheroids, sulfate minerals and kerogen. The Bartle Member has evaporite pseudomorphs and relicts found in close association with the volcanic rocks of the Killara Formation and were pervasively silicified. The Killara Formation volcanic rocks were erupted in a rift setting. Accordingly, the most probable current equivalent for the depositional setting appears to be rift-valley alkaline lakes. This environment is typified by the current Afar region, Northeast Africa, and some of the East African Rift Valley alkaline lakes, e.g., Lake Magadi. The key facts mutual to the Bartle Member and the Afar region are the occurrence of rift-related mafic volcanism, evaporitic, volcaniclastic, and chemical sediments, and carbonate rocks. Geological and petrographic interpretations, combined with geochemical data, and the spatial association of textures and minerals indicate that the cherts of the Bartle Member were most likely formed in an environment of alkaline playa lakes and thermal springs near the end of mafic volcanism in a rift setting. The origin of silica and the timing of the pervasive silicification is unclear, but alkaline brines have been shown to dissolve large quantities of silica, as shown by the ubiquitous replacement of magadiite layers to form the bedded cherts of Lake Magadi (Eugster, 1969, 1970). The presence of microdubiofossils or putative microfossils necessitate an environment favorable to the growth of microbial communities that permit preservation by fast pervasive silicification. Such an environment can be found in thermal springs. The Bartle Member chert comprises anomalous abundances of gold and barium. The combination of silica, organic material, and cyanobacteria is favorable to gold mineralization in a thermal spring environment (Pirajno and Grey, 2002).

A Permo-Triassic pelite-carbonate rock series (with intercalated metabasitic rocks) in the Cordilleras Beticas, Spain, was metamorphosed during the Alpine metamorphism at high pressures (P_{min} near 18 kbar). The rocks exhibit well-preserved sedimentary structures of evaporites such as pseudomorphs of talc, kyanite—phengite—talc—biotite, and quartz after sulfate minerals (Fig. 1.40), and relicts of baryte, anhydrite, NaCl, and KCl, pointing to a salt—clay mixture of illite, chlorite, talc, and halite as the original rock. Kyanite—phengite—talc—biotite aggregates in pseudomorphs formed in the high pressure stage. An especially interesting rock type of the metaevaporites is rare and probably represents less than 5% of the whole exposed sequence. It contains platy pseudomorphs of kyanite—talc—biotite—phengite—paragonite and secondary chlorite + quartz as well as tabular aggregates of talc with inclusions of baryte and anhydrite. Aggregates of quartz, irregular and platy in shape, with small amounts of carbonate and inclusions of baryte, anhydrite, apatite, talc, and phlogopite, as well as aggregates of plagioclase + apatite + epidote appear frequently in all the lithologies. The size of the tabular quartz pseudomorphs varies from a few mm to several cm and the former crystals were euhedral with a platy or tabular form, often oriented radially around carbonate. The shape as well as the mineralogy of these pseudomorphs indicates that they were formed after sulfate minerals such as baryte and gypsum or anhydrite. Replacement of the sulfate phases by quartz or other minerals could have occurred either at an early stage during diagenesis or during the metamorphic evolution. In any case, the well-preserved shape is a strong indication of sulfate minerals, which are typical for an evaporitic environment (Gomez-Pugnaire et al., 1994).

Two types of pseudomorphosed sulfate evaporites were observed in the laminated cherts and interbedded coarse grained carbonates from the SiO_2-dolomite sequence at the Mount Isa mine (Queensland, Australia). Pseudomorphs in the chert layers indicated the original sulfate minerals formed in the host sediment during diagenesis. The silica-dolomites consist of a series of brecciated and recrystallized siliceous dolomites. Early diagenetic gypsum and anhydrite have been replaced by quartz, dolomite, calcite, and pyrrhotite. Rare tiny anhydrite relics have been identified in some dolomite pseudomorphs after gypsum. Likewise, pseudomorphed sulfate evaporites have been found in the McArthur Group (1600–1500 Myr BP) and also in the Paradise Creek Formation (1600–1500 Myr BP). The first type of pseudomorph was most straightforwardly recognized in the laminated chert layers where they cut across and distort the bedding indicating that the original sulfate minerals crystallized in the host sediment before compaction during diagenesis. The lath-shaped pseudomorphs were up to 4 cm long, with a length/width ratio of 10:1 or larger with square cross-sections. This habit is typical of diagenetic anhydrite.

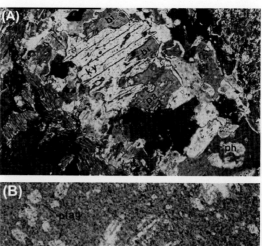

FIGURE 1.40   (A) Photomicrograph: kyanite, ky, and phengite, ph, which occur in tabular pseudomorphs after sulfate minerals, are replaced by biotite, bi, and paragonite, pa. Inclusions in kyanite and talc are baryte and anhydrite (scale bar 0.2 mm, crossed nicols); (B) photomicrograph of a tabular, polycrystalline quartz pseudomorph, qtz, with a "bird tail shape," in a biotite matrix together with plagioclase, plag, blasts (scale bar 0.2 mm, plain polarized light) (Gomez-Pugnaire et al., 1994).

These pseudomorphs now entail a mosaic of quartz grains of a size (50–100 μm) that is constantly larger than that of the enclosing chert. Such textures were thought to indicate that the pseudomorphs have undergone a void phase. The second type of pseudomorphs consist of large (0.1–3 cm) single crystal dolomite replacements after gypsum and were observed in the fine-grained (5–25 μm) dolomitic siltstones and in the coarse-grained carbonates. These pseudomorphed crystals had either angular or diskoidal terminations. The diskoidal types had a similar habit to gypsum that has formed interstitially in recent and ancient sabkha environments. The angular pseudomorphs after gypsum are six sided, with interfacial angles comparable to those of gypsum elongate parallel to the c axis. Numerous pseudomorphs, however, are regularly fractured and brecciated. In some of the dolomite pseudomorphs after gypsum, small relics (10–20 μm) of anhydrite were observed with the petrological microscope, thus demonstrating a complex replacement history of gypsum–anhydrite–dolomite. The preservation of anhydrite relics, of carbonaceous material, the textures, and the reduced nature of the sediments point to the diskoidal gypsum being of evaporitic origin instead of formed by the degradation of sulfides. Additionally, the Mount Isa evaporites have important metallogenic consequences. Numerous models for the formation of stratiform lead–zinc and copper ores involving sulfur and metal rich evaporitic brines have been proposed. Although a volcanogenic model cannot be unambiguously disproven, the recognition of replaced evaporites at both McArthur River and at Mount Isa raises the likelihood that these deposits may have been formed from metal rich brines of the kind forming at present in evaporitic basins (McClay and Carlile, 1978).

The Late Proterozoic Damara Belt of Namibia grew from an elaborate system of continental rifts in which the basal portion (Nosib Group) of the Damara Sequence was deposited. In the southern rift, located at the southern margin of the Damara Belt, the Nosib Group is characterized by coarse clastic sediments (Kamtsas Formation) and fine-grained partly dolomitic deposits (Duruchaus Formation). Both formations occur, with interfingering relation over almost the total length of the rift. In the Geelkop Dome range, the pelitic–dolomitic Duruchaus Formation comprises, in its upper part, a series of sediments that are characterized by cyclical deposition, high sodium contents, abundant albite pseudomorphs after primary evaporite minerals, and concordant solution and collapse breccias. This 300 m thick sequence has been taken as deposits of an alkali lake or playa complex. The unusually high albite content in some rock types of this and the other more evaporitic facies is a distinctive feature of the Gurumanas Evaporite Member. The albite is usually crypto- to microcrystalline and of very high purity (0.3% CaO, 0.1% K_2O). It cannot be a primary mineral because it replaces carbonates in the submerging lake facies, primary evaporite minerals in the submerging/emerging mud flat facies, and evaporite crusts in the saline crust facies. Three different mechanisms of albite formation have been discussed by Behr et al. (1983): (1) Formation of albite from clay minerals during metamorphism to middle green-schist facies. For the crystallization of albite, in the presence of high saline pore fluids that disseminated all through the rocks during metamorphism the following reaction is envisaged (Kulke, 1978):

$$Illite + NaCl(aq) + SiO_2(aq)^- \rightarrow NaAlSi_3O_8 + HCl(aq) + H_2O + KCl(aq) \tag{1.26}$$

It is presumed that throughout this reaction carbonates buffered the liberated HCl and in so doing were replaced by albite. This type of albitization would be limited to pelitic sediments that specially occur in the submergent lake facies. (2) Formation of albite from primary evaporite minerals during early diagenesis. It has been previously recognized that primary gaylussite is transformed to pirssonite and finally to shortite by dehydration reactions during diagenesis. According to Bradley and Eugster (1969) the formation of shortite from pirssonite and calcite occurs with decreasing H_2O at 90°C, whereas with increasing temperature, in an alkaline environment, shortite transforms to albite by the reaction:

$$(Na_2CO_3 \cdot 2CaCO_3) + 2Al_3OOH(aq) + 6SiO_2(aq) \rightarrow 2NaAlSi_3O_8 + CaCO_3 + H_2O + CO_2 \tag{1.27}$$

This type of albite formation is the most probable to describe the extensive occurrence of albite pseudomorphs after evaporite minerals in the submergent/emergent mud flat facies. Laminated sandy albitic dolomite: The most distinctive features of this subfacies are its bright, reddish-brown color, the manifestation of concordant breccias, and the existence of ample idiomorphic albite pseudomorphs, up to 5 cm in size, after several initial diagenetic evaporitic precursors (Fig. 1.41). The sediment matrix is primarily albite and dolomite with further detrital quartz and biotite and metamorphic muscovite. Albite pseudomorphs are observed all through the sandy dolomite, but are concentrated in layers where they cover 50%–80% of the total rock. Mineral habit and sedimentary structures suggest a high evaporitic milieu and indicate essentially sodium carbonates as the precursor minerals of the albite pseudomorphs. All the pseudomorphs in the sandy albitic dolomite consist of crypto- to microcrystalline albite and small amounts of dolomite, calcite, tourmaline, rutile, paragonite, or chlorite, the individual composition contingent on the type of precursor mineral. Several types of pseudomorphs can be identified; these generally fall into one of two size groups: 1–5 cm across or less than 5 mm across. The large pseudomorphs that are more easily identifiable include: (1) Irregularly intergrown clusters of prismatic crystals up to 5 cm in size with a simple habit, and an almost square section, which in places is rhombohedral or wedge like. Skeletal crystal growth with inclusions of matrix components often occurs. Displacive authigenic growth from pore fluids in the soft sediment is inferred by the association between the crystals and the deformed sag structures in the laminated sandy dolomite. Crystal morphology and assessments of recent minerals from Lake Natron (Tanzania) point to the most likely precursor minerals: gaylussite ($Na_2Ca(CO_3)_2 \cdot 5H_2O$), thermonatrite ($Na_2CO_3 \cdot H_2O$) or shortite ($Na_2CO_3 \cdot 2CaCO_3$) (Behr et al., 1983). (2) Large pseudomorphs of albite with a lot of dark green xenoblastic tourmaline inclusions. The tourmaline is limited to the pseudomorphs and has not been observed outside the host rock. The pseudomorphs are monoclinic showing a simple habit of (100), (010), and (001) planes. Skeletal crystal growth, fan-like sections and irregular intergrowths are distinctive features. Although pseudomorphs of this type are not without difficulty distinguished macroscopically from those in (1), their high tourmaline content suggests an origin from different precursor minerals, most probably from hydrated borates, e.g., borax ($Na_2B_4O_7 \cdot 10H_2O$). (3) A less regular but very typical pseudomorph of albite forms cross-twin-crystals at an angle of roughly 100 degrees to each other and up to 1 cm in size. This form of twinning is common in thenardite (Na_2SO_4). Due to poor conservation of crystal habit the smaller pseudomorphs (<5 mm) are hard to identify. There are possibly six different types: only three are cited here. Based on chemical/mineralogical observations the precursor minerals of two of these are tentatively identified. (1) Globular aggregates of weakly zoned albite and dolomite. The primary mineral may have been northupite ($Na_2CO_3 \cdot MgCO_3 \cdot NaCl$). (2) Small ovoid albite pseudomorphs, 2×4 mm in size, with abundant tourmaline and rutile. The primary mineral, in this case, is assumed to have been leucosphenite ($CaBaNa_3BTi_3Si_9O_{29}$). (3) An undetermined pseudomorph that also forms globular aggregates and is described by a high content of xenoblastic tourmaline concentrated in circular sectors. The late diagenetic origin of the precursor mineral, possibly of the borate

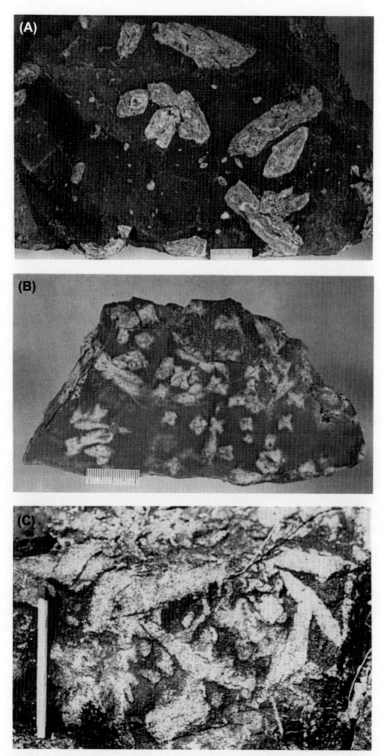

FIGURE 1.41   (A) Clusters of rhombohedral pseudomorphs of albite after gaylussite or shortite in sandy albitic dolomite. (B) Interpenetration twins of thenardite, pseudomorphed by albite, in sandy albitic dolomite. Angle between twins is 102°02″: apparent reduction of angle depending on orientation of crystals in relation to surface of sample. (C) Cluster of tabular trona, pseudomorphed by albite, in sandy albitic dolomite (Porada and Behr, 1988).

group, is inferred by the undisturbed bedding through the pseudomorph. Infrequently large tabular pseudomorphs of albite, up to 4 cm in size, radiating from "albitolite" layers and partially forming rosette-like clusters are developed. They are understood to be former crystals of trona ($NaCO_3 \cdot NaHCO_3 \cdot 2H_2O$). A reconstruction of the geochemical conditions during deposition of the primary minerals in the lake and mud flat domains is difficult and remains partly speculative. It has to be based on the identification of primary and early diagenetic minerals that have been replaced by albite. As the primary mineralogical record has suffered substantial change during successive metamorphic, metasomatic and hydrothermal events, neogenic minerals must also be taken into account. Last, some of the chemical components are trapped in several generations of fluid inclusions. From the description of the sedimentary facies it might be concluded that the evaporitic sequence of the Duruchaus Formation is described by carbonate-dominated precipitates of mostly dolomitic composition, various efflorescences and a number of authigenic primary diagenetic evaporative sodium carbonate minerals that later have been pseudomorphed mainly by albite. Primary evaporite minerals (besides dolomite) so far identified or put forward, are

shortite ($Na_2CO_3 \cdot 2CaCO_3$)
or gaylussite ($Na_2Ca(CO_3)_2 \cdot 5H_2O$)
or thermonatrite ($Na_2CO_3 \cdot H_2O$);
trona ($Na_2CO_3 \cdot NaHCO_3 \cdot 2H_2O$);
thenardite (Na_2SO_4);
borax ($Na_2B_4O_7 \cdot 10H_2O$);
northupite ($Na_2CO_3 \text{-} MgCO_3 \cdot NaCl$);
leucosphenite ($CaBaNa_3BTi_3Si_9O_{29}$).

All fluid inclusions were shown to be highly saline with a maximum salinity of 38 wt% equiv. NaCl. The solutions are NaCl-dominated, but the low eutectic temperatures, up to $-75°C$, indicate the presence of Ca^{2+} and Mg^{2+}, as well as dissolved hydrocarbons and sulfur (Porada and Behr, 1988).

The stratiform Cu−Co ore mineralization in the Katangan Copperbelt comprises dispersed sulfides and sulfides in nodules and lenses, which are frequently pseudomorphs after evaporites. Two types of pseudomorphs can be recognized in the nodules and lenses. In type 1 examples, dolomite precipitated first and was then replaced by Cu−Co sulfides and authigenic quartz, whereas in type 2 examples, authigenic quartz and Cu−Co sulfides precipitated preceding the dolomite and are coarse grained. The sulfur isotopic composition of the copper-cobalt sulfides in the type 1 pseudomorphs was between $-10.3‰$ and $3.1‰$ relative to the Vienna Canyon Diablo Troilite (VCDT), showing that the sulfide constituent was the result of bacterial sulfate reduction (BSR). The production of HCO^{3-} throughout this process resulted in the precipitation and replacement of anhydrite by dolomite. Another product of BSR is the generation of H_2S, resulting in the precipitation of Cu−Co sulfides from the mineralizing fluids. Early sulfide precipitation happened along the rim of the pseudomorphs and continued to the core. Precipitation of authigenic quartz was almost certainly induced by a pH drop during the sulfide precipitation. Fluid inclusion results from quartz suggest the existence of a high-salinity (8−18 equiv. wt% NaCl) fluid, perhaps derived from evaporated seawater that moved through the deep subsurface. Pseudomorphs after anhydrite are plentiful in the Mines Subgroup. They are found in oval, round and cauliflower-shaped nodules, in lenses that might be wedge-shaped and as laths. Two major types of replacement and cementation of these structures can be recognized. In type 1 dolomite crystallized first and was later substituted by Cu−Co sulfides and authigenic quartz. Relicts of anhydrite and halite are still existent in these type 1 nodules and lenses. In type 2 the authigenic quartz and sulfides precipitated preceding the dolomite and the crystals are coarse-grained and free growing. Primary fluid inclusions in type 1 authigenic quartz belong to the $H_2O−NaCl$ system, with a salinity between 8.4 and 18.4 eq. wt% NaCl and homogenization temperatures between 80 and $192°C$. Primary fluid inclusions in type 2 authigenic quartz have high-salinity (38.6−46.5 eq. wt% NaCl) fluid inclusions with homogenization temperatures between 324 and $419°C$. This fluid replicates the mineralizing settings during deep burial and orogenesis. The study by Muchez et al. (2008) focused on the features and formation circumstances of type 1 nodules and lenses to get a better insight in the processes that ensued during diagenesis. Oval to rounded nodules vary in size from a few hundred micrometers to 5 mm. The nodules contain a rim of authigenic quartz and sulfides such as chalcopyrite, bornite, carrolite, and chalcocite. The size of the authigenic quartz crystals varies from 20 to 200 μm. The center of the nodules consist of a brown luminescent hypidiotopic dolomite (up to 1.2 mm). Authigenic quartz and Cu−Co sulfides replaced dolomite crystals and thus postdate dolomite formation. Even though sulfides and authigenic quartz mainly occur around the rim of the nodules and the dolomite in the center, the nodules can also be totally replaced by authigenic quartz and sulfides with rare relics of dolomite. Pseudomorphs after anhydrite were also observed as lenses (size ranging from 250 μm to 1 mm in width and from 1.5 mm to 2.5 cm in length). Wedge-shaped lenses are several centimeters long and up to 15 mm thick. The lenses are composed of authigenic quartz, bornite, carrolite, chalcopyrite, chalcocite, and a few relics of brown luminescent dolomite, which is replaced by authigenic quartz and sulfides (Fig. 1.42). Quartz frequently has inclusions of Cu—Co sulfides, but can also be overgrown by sulfides. Lath-shaped pseudomorphs after anhydrite are ~400 × 100 μm in size and consist of dolomite. The calculated ΔSO_4—H_2S values for the copper—cobalt sulfides in the evaporite pseudomorphs fell between +14.4‰ and +27.8‰, in line with fractionation values typical of bacterial sulfate reduction. Hence, the replacement of anhydrite by dolomite is understood to be largely related to BSR

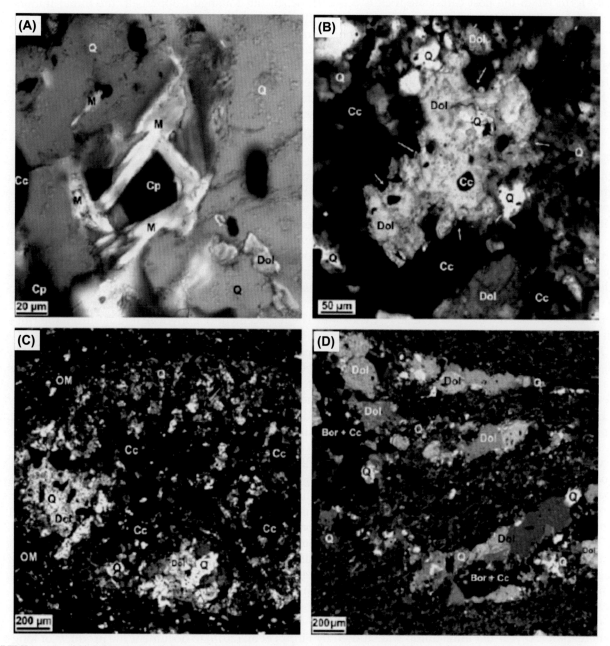

FIGURE 1.42   (A) Muscovite (M) surrounds in a rectangular way chalcopyrite and is enclosed in massive authigenic quartz (Q). *Cp*, chalcopyrite; *Cc*, chalcocite; *Dol*, dolomite; double-polarized light. (B) Pseudomorphic nodule after anhydrite consisting at its rim of authigenic quartz (Q) and chalcocite (Cc). Both minerals replace (*arrows*) dolomite (Dol) which is still present in the core of the nodule; double-polarized light. (C) Oval pseudomorph after anhydrite consisting of authigenic quartz (Q) and chalcocite (Cc) with relict dolomite crystals (Dol) in an organic-rich layer; double-polarized light. (D) Wedge-shaped pseudomorphs in dolomitic siltstone consisting of dolomite (Dol), authigenic quartz (Q) and bornite (Bor) and chalcocite (Cc); double-polarized light (Muchez et al., 2008).

occurring in the anhydrite nodules. Nonetheless, the model does not vary even when there is input of sulfide from thermochemical sulfate reduction by hot mineralizing fluids as the sequence of reactions remains the same.

$$\text{Organic matter} + SO_4^{2-} \rightarrow \text{altered organic matter} + H_2S(HS^-) + HCO_3^-(CO_2) + H_2O \qquad (1.28)$$

The reductant may possibly have been provided by organic matter near the nodules and disseminated in the siltstones or by cyanobacterial mats in the laminated carbonates. Calcite frequently replaces calcium sulfate, as proven by the widespread existence of calcite cements in environments affected by BSR. In addition to calcite, dolomite may possibly form when the host carbonate rock is a dolostone. In the Katanga Copperbelt, the initial mineral that replaced anhydrite in the dolomitic siltstone and dolomite was definitely dolomite. An important observation is the close relationship between sulfides and large amounts of authigenic quartz in the pseudomorphs. Crystallization of large quantities of quartz is only conceivable if adequate volumes of a silica-rich fluid migrated through the siltstones and dolomites. Si-rich fluids are characterized by an elevated pH. The substantial volume of copper–cobalt sulfides in the pseudomorphs suggests an open to semiopen system for the mineralizing fluids, permitting effective mass transfer of the metals. The saline, metal-bearing fluids moved through the sediments and, in the vicinity of the pseudomorphs, reacted with H_2S produced during BSR, resulting in the precipitation of sulfides:

$$Me^{2+} + H_2S \rightarrow Me_S + 2H^+ \qquad (1.29)$$

Precipitation initially transpired around the rim of the pseudomorphs and continued toward the center. Occasionally the core still contains anhydrite. Throughout the precipitation of the sulfides, H_+ ions were released that are understood to have resulted in a decrease in pH and crystallization of authigenic quartz. Brown (2005) previously proposed that the mineralizing fluid in sediment-hosted stratiform copper deposits might be buffered by silicate constituents of the basin fill. The solubility of quartz increases considerably with pH. A pH drop could also result in the dissolution of previously formed dolomite in the pseudomorphs and of the more fine-grained dolomite in the host rock. As BSR occurs at temperatures from 0 up to about 80°C, this reaction did not take place during the infiltration of the hot mineralizing fluid, with temperatures up to 190°C. Although, in general, authigenic quartz and sulfides replace dolomite, dissimilar consecutive phases can be recognized and the quantities of the three minerals can vary considerably within the nodules between different layers. Nevertheless, within one layer, the mineralogical composition of the pseudomorphs is relatively constant. The replacement of anhydrite nodules from the rim to the core is representative of these pseudomorphs and suggests that transport toward the core remained possible. Typical is the high amount of carrolite in the rocks that are rich in pseudomorphs after anhydrite. This may be elucidated by the high sulfur activity in the host rock due to BSR. Craig et al. (1979) showed that the type of mineral species that crystallizes in the Cu–Co–S system mostly hinges on the temperature and sulfur activity. Carrolite specifically crystallizes at very high sulfur activities, in accordance with the large volumes of sulfur accessible in the anhydrite nodules. Existing data and this study (Muchez et al., 2008) indicate that organic carbon played a significant part in the precipitation of dolomite, but there is no benchmark to differentiate early from late oxidation reactions. The wide range of measured oxygen isotope values might be due to a difference in the precipitation temperature of the dolomite, variation in the isotopic composition of the dolomitizing fluid, variable interaction of the ambient fluid with the host rock or a combination of these processes. The oxygen isotope data of dolomite in the type 1 nodules, layers, and bands at Kamoto only show that if dolomite precipitation happened from sea water, the calculated temperature of the dolomitizing fluid was between 55 and 70°C using the fractionation factors published by O'Neil et al. (1969), and therefore within the range of bacterial sulfate reduction. Based on the negative $\delta^{34}S$ values of $-10.3‰$ to $+3.1‰$ VCDT of the metal sulfides in the type 1 evaporite pseudomorphs, the replacement of the type 1 nodules and lenses was inferred to be connected with bacterial reduction of sulfate resulting from primary anhydrite. Production of CO_2 during this reaction initiated the precipitation of dolomite. When the mineralizing fluid reached the BSR reaction site, the H_2S reacted with the metals Cu, Co, and Fe to form sulfides such as carrolite, bornite, chalcopyrite and chalcocite. During sulfide precipitation, H^+ ions were released, resulting in a decrease in pH and crystallization of authigenic quartz (Muchez et al., 2008).

A more than 250-year-old mine dump was studied by Filippi et al. (2015) to document the products of long-term arsenopyrite oxidation under natural conditions in a coarse-grained mine waste dump. A discrete mineralogical zonation was observed (listed based on the distance from the decomposed arsenopyrite): (1) scorodite (locally associated with native sulfur pseudomorphs) plus amorphous ferric arsenate (AFA/pitticite), (2) kaňkite, (3) As-bearing ferric (hydr)oxides and jarosite (Fig. 1.43). Arsenopyrite crystals and aggregates enclosed in altered rock were often replaced ("pseudomorphed") by native sulfur [S], which formed vermicular aggregates consisting of imperfectly developed microcrystals. Filippi (2004) described S enclosed in scorodite cement in waste ore concentrate. Jeong and Lee (2003) and Courtin-Nomade et al. (2003) documented S pseudomorphs after sulfides enclosed in ferric hydroxides (often mixed with ferric oxides). Basu and Schreiber (2013) observed arsenopyrite grains embedded by S and scorodite in a mineralized host rock, whereas Majzlan et al. (2014a,b) found S pseudomorphs after argentopyrite

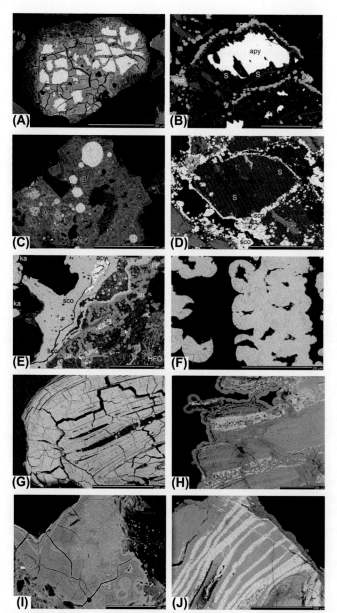

in massive As. It appears that formation of the S pseudomorphs is a quite common feature for arsenopyrite, which is encapsulated by secondary phases or grown in a host rock, i.e., somehow separated from the adjacent conditions. Nesbitt and Muir (1998) indicated that diffusion of As to the oxidized surface supported rapid, selective leaching of As, leaving behind an S-enriched layer. McGuire et al. (2001) suggested a multistep oxidation reaction in which S forms at the arsenopyrite surface via a series of sulfoxy anions ending in thiosulfate, which is released into solution. Later, due to the acidic environment, thiosulfate is altered to bisulfite and elemental S. Nevertheless, this process does not explain the massive S accumulation. The laboratory experiments of Asta et al. (2010) proved that arsenopyrite dissolution is severely influenced by dissolved oxygen and only slightly affected by the pH and temperature. They also indicated that exposed arsenopyrite was covered by S-enriched surface layers under acidic conditions, whereas Fe-rich layers dominated under mildly acidic to basic pH conditions. Formation of S with arsenopyrite here, as well as in the work by Filippi (2004), Jeong and Lee (2003), and Courtin-Nomade et al. (2003), seems to progress under restricted access of oxygen when the sulfide relicts are encapsulated by secondary products. According to Ondruš et al. (1997) slow decomposition of arsenopyrite in cement and the catalytic role of the H_2SO_4 present result in the surplus arsenate acid being gradually removed or neutralized and it may locally result in higher pH conditions.

A study of Co mineralization in weathered ultramafic rocks of New Caledonia revealed the presence of heterogenite (2H and 3R polytypes), asbolan (with Co and Ni), lithiophorite, intermediate phases between asbolan and lithiophorite, cryptomelane, ramsdellite, and todorokite (Llorca and Monchoux, 1991). Cryptomelane, ramsdellite, and todorokite contained merely traces of Co. The chief Co phases were found not only as cryptocrystalline aggregates but also as crystals or fibers up to several hundred microns in length. They form concretions, but more frequently as pseudomorphs after silicates and plant roots, which are habits not before recognized for these minerals. The pseudomorphs after the silicates (serpentine, olivine, and talc derived from pyroxene in the initial stages of weathering) were observed along joints, resulting in a macroscopic appearance of black spots or coatings. Cobalt minerals ordinarily replace veinlets of serpentine. In the second, more common type of replacement of the serpentine, single crystals are formed; perfectly replicating the serpentine veinlets. Ni-rich asbolan and Ni-rich "asbolan–lithiophorite intermediates" generally occur in this habit. Synchronous pseudomorphs after talc crystals were frequently observed for the prominent minerals (asbolan, Ni-rich in particular, "asbolan–lithiophorite intermediates",

FIGURE 1.43 Back-scattered electron (BSE) images showing different relationships among the selected As-bearing minerals: (A) relicts of arsenopyrite crystals (white color) replaced by cracked AFA/pitticite; (B) relict of arsenopyrite crystal (apy) replaced by sulfur (S) and rimmed by globular scorodite (sco) aggregates (gray color); (C) spherical–framboidal aggregates of pyrite in a carbonate; (D) sulfur pseudomorph preserving the shape of the arsenopyrite crystal and rimmed by scorodite globular aggregates; (E) partly decomposed arsenopyrite crystals enclosed in scorodite matter (sco) and chemically zoned jarosite crystals (jar), the heterogeneous matter in the bottom right-hand corner consists of HFO (ferric hydroxides often mixed with ferric oxides) and rock-forming silicates, kaňkite aggregates are the top left corner (ka); (F) kaňkite spherical–globular aggregates; (G) laminated AFA/pitticite; images (H–J) show some of the complex and structurally different ferric (hydr)oxides, the bright parts represent the predominance of hematite and the dark parts the predominance of goethite (Filippi et al., 2015).

heterogenite, lithiophorite). Crystallinity of the newly formed phases was consistently good, due to oriented growth of the fibers from and perpendicular to the cleavage planes and faces of the dissolving talc crystals. Cavities between the serpentine veinlets, formerly occupied by olivine cores, are in some instances packed with a disordered crypto-crystalline cobaltiferous material. These pseudomorphs, composed mostly of heterogenite or Ni-rich asbolan, were formed asynchronous or subsynchronous with the dissolution of olivine. They were less frequently observed than the pseudomorphs after talc or serpentine. Fossilized roots are common. Microscopic and macroscopic examinations indicate that the root was first cast by an optically isotropic cryptocrystalline mass, which can even enter the pores and fill the inner wall while the plant is alive. The tissues themselves are then more or less totally replaced, occasionally by an optically anisotropic material of the same nature, ordinarily "asbolan-lithiophorite intermediates," but Co-rich asbolan or lithiophorite as well. Specifically, around the roots, where the porosity at the level of individual cells, is high, cobalt minerals can impregnate the earthy mass. Microscopic analysis of such impregnations displayed an isotropic and cryptocrystalline material with irregular or diffuse contours. The same minerals were observed here as in fossilized roots. Cobaltiferous fibers, mainly made of asbolan, and showing a propensity to spherulitic arrangement, were observed to have formed in cavities left by the dissolution of rock-forming silicates (talc or olivine?). In this situation, the pseudomorphs were subsynchronous with the dissolution, and only the external shape of the mineral was preserved.

The composition and microstructure of opalized saurian bones (Plesiosaur) from Andamooka, South Australia, were analyzed and compared to saurian bones that have been partially replaced by magnesian calcite from the same geological formation, north of Coober Pedy, South Australia by Pewkliang et al. (2008). During the formation of the opal, the cruder details of the bone microstructure have been preserved down to the level of the individual osteons (scale of around 100 μm), but the central canals and the boundary area have been enlarged and filled with chalcedony, which postdates opal creation (Fig. 1.44). These chemical and microstructural characteristics are in line with the opal formation being a secondary replacement after incomplete replacement of the bone by magnesian calcite. They are also in accordance with the opal forming first as a gel in the small voids left by the osteons, and the separate opal spheres growing as they settle within the gel. Variations in the viscosity of the gel offer a complete reason for the occurrence of color and potch banding in opals. The suggestion that opalization is a secondary development after calcification on the Australian opal fields is in keeping with a tertiary age for formation. The process of mineralization of bone and shell for conservation in the fossil record is reliant on the chemical and physical conditions during diagenesis, predominantly the composition of the mineralizing fluid. Essentially, these processes may be seen as mineral—replacement reactions, in which the biomineral is replaced by another mineral such as quartz, opal, or calcite. The replacement reaction can also be related to recrystallization of the biomineral, particularly biogenic apatite (carbonate-hydroxylapatite), through which trace amounts of heavy metals can be assimilated into the apatite structure. For the very-fine-scale microstructure of the fossil to be well-preserved, the dissolution reaction that affects the biomineral and any existing soft tissue need be closely coupled to the precipitation reaction, so that little free space forms at the reaction front (Putnis, 2002). There are two types of opal: microcrystalline opal (consisting of opal-CT (tridymite) and opal-C (cristobalite)) and noncrystalline opal (consisting of opal-A, opal-AN, and opal-AG). Opal-A is a biogenic form of silica, opal-AN ("hyalite") is a hydrous amorphous silica glass, opal-AG is the form found in precious opal (which shows a play of colors) and potch opal (which does not show a play of colors). Opal-AG is an ordered arrangement of spheres (1700—3500 Å in diameter) comprising noncrystalline, amorphous, gel-like silica with H_2O filling the interstices. Australia's precious opal is found in veins and nodules in bentonite clay layers in shales and sandstones and has no direct relationship with magmatic activity (i.e., it is "sedimentary" opal). Precious and potch opal have also been observed replacing shells, wood and, less frequently, the bones and teeth of marine reptiles. The skeletons of Cretaceous plesiosaurs and ichthyosaurs that have been in part or totally opalized have been collected at Andamooka, Coober Pedy, and White Cliffs, Australia. The microstructure of the fossilized limb bone of an ichthyosaur from Moon Plain north of Coober Pedy is similar to that of the modern dolphin bone. The fibrous rings of biogenic apatite can just be observed accompanied by the black canaliculi, even though a carbonate mineral has substituted the collagen. The central canals seem marginally enlarged and are occupied with magnesian calcite. The margins between the individual osteons can also be somewhat enlarged and filled with magnesian calcite. Largely, the calcification of the bone has well-preserved details of the microstructure on a micrometer scale. The microstructure of the opalized plesiosaur bone is identifiable as bone and the outlines of the individual osteons are obviously visible. The central osteone canals and the boundary areas have been enlarged by a factor of no less than two in comparison to recent dolphin bone, and no longer have smooth, sharply defined edges. These areas are occupied by fibrous chalcedony, as are cracks joining some of the canals to the boundary areas. The biogenic apatite has been entirely replaced by opal, which has a slightly cloudy look due to the

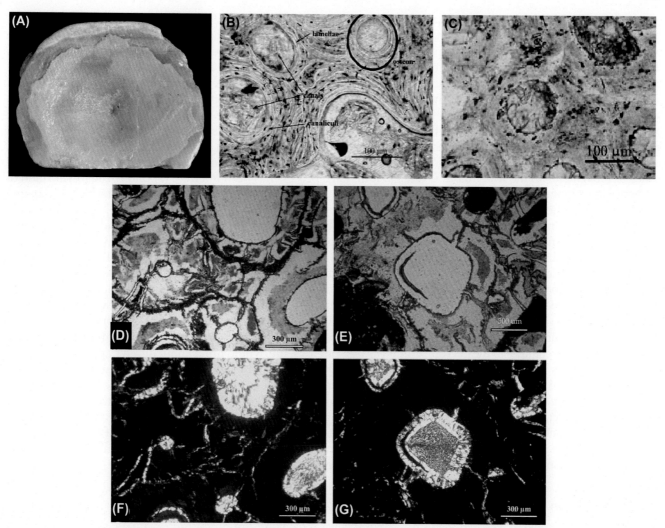

FIGURE 1.44 (A) Opalized vertebra from ichthyosaur *Platypterygins longmani* from the Bulldog shale, Coober Pedy, South Australia (on loan to the South Australian Museum). The vertebra is 5.8 cm across. Note the horizontal color banding in the opal. (B) Microtexture of rib bone from a modern dolphin (*Delphinus delphis*). The ostenos, Haversian canals, canaliculi, and the lamellae of biogenic apatite and collagen are indicated. (C) Microtexture of a fossilized bone of an ichthyosaur from Moon Plain, north of Coober Pedy, South Australia. Note the degree of microstructural preservation, indicating a closely coupled replacement-reaction during fossilization. (D−G) Photomicrographs of the microstructure of opalized plesiosaur bone from Coober Pedy. Note that only the outlines of the osteons and the Haversian canals are preserved, not the fine microstructure as seen in the fossilized bone (C). The canals and the spaces around the osteons are filled with chalcedony. Images (D) and (E) are taken in plane-polarized light, and images (F) and (G) are taken with crossed nicols (Pewkliang et al., 2008).

inclusion of micrometer size scales of kaolin. The canaliculi have fully disappeared, as has the fibrous microstructure of the osteons. The scale of the microstructural details still visible is around 100−200 mm. The opalized bones are principally pure SiO_2, with negligible amounts of Al_2O_3. The opalized bone samples have no relict biogenic apatite. This is in line with the bone structure having been totally recrystallized and replaced by opal and silica. The Haversian system is preserved in the fossilized ichthyosaur bones, even though the central canals and the boundary regions have become enlarged through replacement by magnesian calcite. The carbonate-rich fluids therefore moved through these openings throughout diagenesis and then infiltrated the ostonal structure, substituting the collagen and other organic material and recrystallizing the biogenic apatite. The biogenic apatite recrystallized during diagenesis, and the neighboring collagen and further organic material were replaced by carbonate. This micrometer-scale preservation of the bone structure suggests that the dissolution and replacement reactions were closely coupled. At the end, the canals and boundary veins were filled in with magnesian calcite. The enrichment in Sr indicates that this happened in a marine environment, maybe shortly after burial. In contrast, the opalized bones only exhibit outlines of the Haversian system. The canals and boundary veins are significantly enlarged, have irregular boundaries and are occupied by fibrous chalcedony. The chalcedony evidently postdates opal formation, as it fills cracks and surrounds small slithers of opal at the edges of the canals. The opal has totally replaced the osteon structure, leaving no biogenic apatite or the fibrous microstructure. The reactions that resulted in the dissolution of the biogenic apatite and the formation of opal hence were possibly not closely coupled, and the opal formed in free space. Furthermore, the cracks around the osteon margins and from the central canals suggest that the opal shrunk as it solidified. The nature of the conservation of the bone microstructure leads to a number of questions about the formation of the opal. From the structural detail, it is clear that the opal did not fill an open cast of the bone in the sediments. The observation that the boundary cracks and the canals are preserved and became enlarged indicates that they were previously filled with a mineral before the opalization reaction. The mineral was afterward replaced by chalcedony after the opal had solidified, conceivably as the consequence of a change in pH of the groundwater or movements in the water table. Overall, a four-stage process has been inferred: (1) diagenetic fossilization of the bones in a manner comparable to those still preserved at Moon Plain, with recrystallization of the biogenic apatite and infill of the Havesian system by a carbonate mineral; (2) dissolution of the biogenic apatite, generating small ($\sim$100 mm) voids in which opal and kaolinite are later formed; (3) dissolution of the carbonate, and (4) precipitation of chalcedony in canals and cracks. Processes (3) and (4) appear to have been coupled, and might have been helped by the shrinkage resulting from opal "solidification", which could have allowed fluid flow into the fossil bone. The microstructure of the chalcedony visibly exhibits a series of deposition episodes. The size of free space in some of the smaller osteon cavities is about 100 µm, equivalent to the diameter of around 500 opal spheres. The indication of shrinkage in the osteon-filling opal strongly indicates that the growth of the opal spheres happened within a silica gel that filled the network of osteons. The opal spheres grew as they gradually settled in the gel, conceivably by a process of Ostwald ripening. Fluctuations in the viscosity of the gel, caused by oscillations in the level of the water table, from variations in climate or just from changes in the gel composition as the spheres settle, would bring about different periods of growth of the spheres, thus providing a ready explanation for the layering found in some opalized bone and opal. The suggestion that the opal spheres grew in a gel is similarly reinforced by the occurrence of monomineralic, well-crystallized authigenic kaolinite, as a larger diversity of clay minerals and of morphologies would be anticipated if this kaolinite signified a detrital component picked up by the opal-forming fluid. Note that this process does not entirely exclude the probability that nucleation of opal was started in the sediment profile. But it seems probable that if the silica spheres were formed in the profile, they must have been gel-like and contained some water, and they must have dried out during solidification, causing the shrinkage observed in the microstructure (Pewkliang et al., 2008).

Pseudomorphs in Magmatic Settings

Al-Shanti et al. (1984), in their study on the geochemistry, petrology, and Rb-Sr dating of the Jabal Tays trondhjemite and granophyre that intrude the melange zone of an ophiolitic complex, reported the presence of chlorite occurring as scaly aggregates and seems to be a pseudomorph after preexisting mica, amphibole, or pyroxene. Within the group of the mafic minerals they observed epidote pseudomorphs after hornblende. The Jabal Tays trondhjemites were probably produced by partial melting of basic material at the root of an island arc over a subduction zone. magma generation by partial melting of a preexisting lower crust with a short time residence. The granophyric rocks probably originated by crystal fractionation of the trondhjemitic liquids or by contamination of such liquids with the acidic host rocks.

Incipient-stage alteration products in reasonably fresh oceanic gabbros from deep boreholes can provide critical data on hydration processes in the oceanic lower crust and their influence on lithosphere dynamics. Nozaka and Fryer (2011) presented the results of a petrographic study on the alteration of olivine-bearing gabbroic rocks recovered from the deeper parts of Integrated Ocean Drilling Program (IODP) Hole U1309D in the Atlantis Massif near the Mid-Atlantic Ridge at 30°N. In these rocks, alteration is contained to the vicinity of fluid-infiltration veins or igneous contacts. It is most noticeable in halos around amphibole + chlorite veins or leucocratic veins in olivine-bearing gabbros, where coronitic fringes of tremolite, chlorite and talc are found around discrete olivine grains (Fig. 1.45). Numerous halos show a zonal pattern with regular changes in mineral assemblage, commonly comprising three zones: tremolite + chlorite around relict olivine-plagioclase contacts; talc pseudomorphs after olivine; and tremolite pseudomorphs after olivine. They established that the zoned halos were initiated by metasomatism due to protracted or sequential infiltration of hydrothermal fluids at amphibolite-facies conditions (450–750°C, 1.5–2 kbar). Textural relationships obviously suggest that zoned halos formed before the serpentinization and clay mineral formation, and indicate that the high-temperature, amphibolite-facies alteration occurred in a near-axis region before the exhumation of the lower crustal rocks. Microscopic examination of typical rocks have shown that a lot of the halos are mineralogically zoned (zoned halos). The halos contain three zones: a relict olivine zone; a talc zone; and

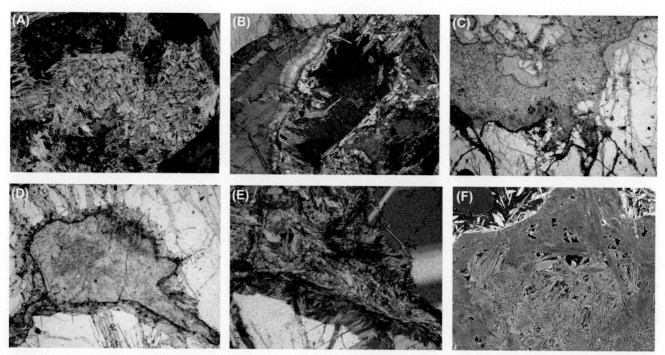

FIGURE 1.45   (A) Enlarged photomicrograph of a part of the Tremolite Pseudomorph (Psd.) Zone (crossed polars). Tremolite aggregates form pseudomorphs after olivine in close association with thick fringes of chlorite that surround adjacent plagioclase. Some of the tremolite crystals are overgrown by later green hornblende. (B) Enlarged photomicrograph of a part of the Relict Olivine Zone (crossed polars). The tremolite–chlorite fringes surrounding olivine and plagioclase are thinner in this zone than those in the other zones. (C) Enlarged photomicrograph (plane-polarized light). Microcracks filled with chlorite and/or green hornblende are developed particularly in the plagioclase of the coarse-grained gabbro. Green hornblende is abundant at the rims of chlorite coronas in contact with amphibole-filled cracks, whereas it is minor (but not absent) in coronas in contact with chlorite-filled cracks. (D) Photomicrograph of green hornblende that cuts coronitic chlorite and fringes a pseudomorphic tremolite aggregate after olivine (plane-polarized light). (E) Enlarged photomicrograph of a part of (D) (crossed polars with a gypsum plate). Green hornblende clearly cuts coronitic chlorite that consists of oriented fibrous crystals shown by a yellow interference color. (F) Back-scattered electron image of a part of (D). Single tremolite crystals are fringed with brighter rims, which have more aluminous, hornblendic compositions (Nozaka and Fryer, 2011).

a tremolite pseudomorph zone. Although some of the "tremolite" crystals are actually Mg-rich actinolite in composition or at the transition between tremolite and actinolite, this series of amphiboles is denoted from now as tremolite for convenience. Tremolite Pseudomorph Zone: The zone closest to the vein or intrusive contact consists of tremolite aggregates that form pseudomorphs after olivine. The replacement of olivine by these aggregates was based on the morphological resemblance to olivine and the close association with coronitic chlorite, which is common around relict olivine grains in weakly altered parts of the rock (the relict olivine zone). The tremolite aggregates most likely formed from the complete breakdown of olivine by the same reaction through which tremolite + chlorite was formed in the relict olivine zone. Relics of primary plagioclase and clinopyroxene persist in this zone, even though they are more intensely replaced by chlorite and amphibole, respectively, than in the other zones. Talc Zone: Talc replacing olivine occurs exclusively in this zone. The replacement is complete and results in pseudomorphs after olivine near the margin with the tremolite pseudomorph zone. Replacement is partial, though, near the boundary with the relict olivine zone. The modal quantity of talc to olivine increases near the veins. Talc pseudomorphs after olivine are fringed by tremolite + chlorite coronas. Where the replacement of olivine is partial, talc crystallizes asymmetrically thicker on the veinward side of olivine crystals. Relict Olivine Zone: The furthest zone from the veins in the halos consists of coronitic tremolite + chlorite fringes around olivine surrounded by plagioclase grains. The coronitic chlorite has a cross-fiber structure, in which minute, fibrous crystals of chlorite are oriented at perpendicular to the contact with plagioclase. Tremolite in this zone not once cuts the fibers of coronitic chlorite, whereas green hornblende in the tremolite pseudomorph zone does. There is no talc in this zone, but the coronitic tremolite and chlorite are also found in the talc zone. In incipiently developed coronas, chlorite is more profuse than tremolite. Even though clinopyroxene in the relict olivine zone appears almost unaltered, tremolite needles grow at the boundary between clinopyroxene and olivine. Their close connection with and parallelism to the veins indicate that zoned halos are initiated by hydrothermal fluids passing through fractures in gabbroic rocks. The chief problem is the source of the regular zonal pattern of the halos. The key to unraveling this appears to be the prominent contrast in distribution and type of occurrence between talc and the tremolite + chlorite assemblage. Talc is irregularly distributed, limited to the talc zone, and exhibits an asymmetric growth pattern on the olivine grains around which tremolite + chlorite exhibits a symmetric, coronitic texture. These textures cannot be described by a single episode of metasomatism, and indicate a difference in the timing of crystallization of the talc and the tremolite + chlorite. It has been frequently observed in natural and experimental systems that Al is virtually immobile in hydrothermal alterations of basaltic or gabbroic rocks under greenschist to amphibolite—facies settings. In the gabbros from this study, the immobility of Al is also indicated by the fact that chlorite is constrained to the vicinity of plagioclase. Additionally, the abundance of the relict minerals left behind in the zoned halos points to a low water—rock ratio during the hydrolysis reactions. When the water—rock ratio is low, Mg precipitates as hydrous magnesian minerals rather than being removed in the hydrothermal fluids. Calcium is solely assimilated into tremolite structure amid the reaction products in the zoned halos. Although Ca can be a mobile element, its addition from outside the system to form tremolite is not required because of the presence of plagioclase as the source of Ca.

$$4CaAl_2Si_2O_8 + 15Mg_2SiO_4 + 5SiO_2(aq) + 18H_2O \rightarrow 2Ca_2Mg_5Si_8O_{22}(OH)_2 + 4Mg_5Al_2Si_3O_{10}(OH)_8$$

$$\text{plagioclase} + \text{olivine} \rightarrow \text{tremolite} + \text{chlorite} \tag{1.30}$$

$$3Mg_2SiO_4 + 5SiO_2(aq) + 2H_2O \rightarrow 2Mg_3Si_4O_{10}(OH)_2$$

$$\text{Olivine} \rightarrow \text{talc} \tag{1.31}$$

Both reactions could occur at T < 600−650°C. At or above 650°C, reaction (1.30) could not occur even at high H_2O activities. A prominent conclusion drawn is that the minerals of the zoned halos could be formed by the reaction of olivine ± plagioclase with fluids of similar compositions at similar T. This seems to be an appropriate condition for the close association of talc with tremolite and chlorite around single veins. Nevertheless, the reactions with fluids of similar composition ought to take place at different times and different silica and/or water activities: i.e. tremolite + chlorite were formed by reaction (1.30) first, and then talc was formed by reaction (1.31) after the end of reaction (1.30) and at higher silica and/or water activities. Actually, the increase in the extent of tremolite + chlorite toward the veins and the presence of the tremolite pseudomorph zone, where olivine is totally decomposed, indicates the increase of water activity and the completion of reaction (1.30) near the veins. Still, talc is under no circumstances found in the tremolite pseudomorph zone but merely in the talc zone, where the existence of relict plagioclase as well as olivine (together with grains altered to talc) indicates that reaction (1.30) did not continue to completion. Talc formation in the talc zone can be elucidated if olivine had previously been separated from plagioclase by tremolite and chlorite via reaction (1.30). Such a reaction sequence is consistent with the difference in the manner of occurrence; i.e., coronitic tremolite + chlorite versus unevenly grown talc. The model for the development of the symmetrical zonal pattern of the alteration halos is that at an initial stage of fluid infiltration through fractures, tremolite + chlorite formed preferentially at the boundary between olivine and plagioclase in the wall rocks. Once this reaction totally decomposed olivine, which is generally secondary in abundance to plagioclase in the gabbroic rocks, water and/or silica activity is raised in the vicinity of the veins. At a later stage, olivine that had been separated from plagioclase by the previously formed tremolite + chlorite was replaced by talc. The amount of silica per olivine molecule required for reaction (1.31) is five times larger as that for reaction (1.30). Because of this highly silica-consuming reaction, the silica activity of the fluid would be rapidly lowered and the amount of talc, and therefore the width of the talc zone, limited. The T condition for reaction (1.31) might be comparable to, or slightly lower than, that for reaction (1.30); fluid flow might be constant or discontinuous after tremolite + chlorite formation. Green hornblende exhibits a mode of occurrence suggestive of its later formation than tremolite + chlorite. The green hornblende is poorer in Si and richer in Al or Tschermakite component than the earlier tremolite. It has been recognized that Ca-amphibole coexisting with an Al-rich phase such as plagioclase and chlorite in metabasites has a higher Al content at higher-T and/or higher-P conditions. Because the green hornblende exhibits an increase of Al substituting for Si in the tetrahedral site, and because there is no reason to assume an increase in P, it is probable that a prograde reaction initiated by an increase of T happened to form the green hornblende after the formation of the tremolite + chlorite coronas. A possible reason for the reaction is heating by the permeation of high-T fluids, because the distribution of green hornblende is concentrated around the veins (tremolite pseudomorph zone).

The concordance of hornblende distribution with the zonal pattern of the halos indicates that higher-T fluids following the same pathway as the earlier fluids that produced the formation of the tremolite + chlorite and talc (Nozaka and Fryer, 2011).

A partially melted pelitic xenolith, found in the Wehr volcano, East Eifel, Germany, contains staurolite porphyroblasts pseudomorphed by a hercynite-rich assemblage that includes ferrogedrite, sillimanite-mullitess, quartz and siliceous, peraluminous glass, which can be schematically exemplified by the reaction, staurolite → hercynite + gedrite + Al-silicate + quartz + melt (Fig. 1.46). Minerals and glass of the pseudomorph are a disequilibrium breakdown assemblage that represents a time temperature transformation of staurolite during short-term heating and cooling. Microstructural evidence points to the hercynite having formed first together with melt. This was followed by ferrogedrite, and crystallization of Al-silicate and quartz from the melt. The lack of cordierite, the occurrence of ferrogedrite (a possible metastable intermediate phase to formation of almandine) in the staurolite breakdown assemblage and confirmation of melting of muscovite + quartz and plagioclase (An20) + quartz in the xenolith point to minimum pressure conditions of 0.35 GPa at ~665−700°C with probable overstepping of the staurolite stability by 40−95°C. Such conditions were kinetically favorable for the creation of the metastable breakdown assemblage that was preserved by quenching on eruption. No staurolite was preserved in the studied sample, but porphyroblastic staurolite analyzed from another Wehr schist xenolith was compositionally homogeneous, Fe-rich ($X_{Fe} \approx 0.79$), and contains ≈ 0.5 wt% ZnO. The formula calculated on the basis of 48(O), four hydroxyls and all iron as FeO was $H_4(Fe,Mn,Mg,Zn)_{4.2}Al_{17.5}Si_{7.8}O_{48}$. The staurolite is accompanied by graphite and ilmenite (as in the xenolith containing pseudomorphed staurolite) and is inferred to contain negligible Fe^{3+}. The recalculated composition falls within the range of Fe^{3+}-poor staurolites. The earliest stage of staurolite breakdown was observed in a buchitic staurolite schist xenolith with the formation of hercynite and melt (glass) along cracks, grain boundaries and some cleavage planes and strongly enclosed contacts between staurolite and glass indicating melting. As a discrete (mass balance) process, this should be strongly degenerate in the Fe−Mg−Al−Si−H−O system (where "O" accounts for redox reactions involving Fe^{2+} and Fe^{3+}) with co-linearity between hercynite−staurolite−melt that requires the melt to be extremely Al-rich. This is not the case. Mass was not conserved and the melt was silica-rich and Al-poor demonstrating open system behavior as inferred from the abundance of glass. In the schist xenolith containing pseudomorphed porphyroblasts of staurolite, hercynite is the foremost reaction product

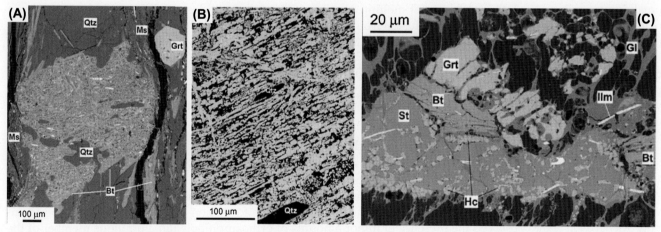

FIGURE 1.46 (A) Back-scattered electron image (BEI) of staurolite porphyroblast extensively replaced by hercynitic spinel in a partly melted politic xenolith, Wehr volcano, East Eifel, Germany. Fine-grained phases in dark interstitial areas to hercynite are not resolved at these magnifications. *Qtz*, quartz; *Ms*, muscovite (partially melted); *Bt*, biotite; *Grt*, garnet. Elongate bright grains in porphyroblast are ilmenite. Light gray areas within biotite are spinel rich. Muscovite is characterized by abundant thin black striations that are glass-lined cleavage planes associated with tiny bright grains of spinel. (B) High-contrast BEI enlargement of part of [7(A)] showing dimensional orientation of hercynite (white), probably parallel to the {010} cleavage of staurolite, and cross-cutting veins of hercynite that presumably represent original cracks in the staurolite porphyroblast. The hercynite is intergrown with gedrite (light gray). (C) BEI showing staurolite (St) breakdown to hercynitic spinel (Hc; white grains) and melt (glass) along cracks in a buchitic politic xenolith, Wehr volcano, East Eifel, Germany. *Grt*, garnet; *Bt*, biotite; *Ilm*, ilmenite; *Gl*, glass. The staurolite exhibits strongly embayed contacts with glass indicating melting. Note that most of the hercynite grains contain inclusions of glass (*darker gray spots*). Biotite shows the initial stage of partial melting with the development of melt-lined cracks that sometimes contain small granules of pleonaste spinel (Bt, lower right side of photo). Although disrupted by glass, almandine-rich garnet (Grt) is unaltered. The web-like pattern of glass around vesicles (black) is the result of H_2O exsolution from the melt when the xenolith was erupted (Grapes and Li, 2010 © www.schweizerbart.de).

(c. 70%). Dimensional orientation of most hercynite in the porphyroblast may have been controlled by expansion along {010} cleavage planes during dehydroxylation of the staurolite with increasing temperature. This would have provided optimal low activation energy sites for the nucleation and growth of hercynite, the kinetically most favorable phase to crystallize with increasing temperature. Reaction along cleavage planes is also likely to have been comparatively fast because nucleation and growth were promoted by rapid diffusion, particularly if melt is present as in this case. Substitution of Si in hercynite of the porphyroblast, particularly in rims that are in contact with glass and inclusions of glass, hint at growing in contact with the melt. Textural relations between gedrite and hercynite suggest that gedrite crystallized later, probably when temperature reached an appropriate level (possibly slight cooling) allowing it to nucleate along interfaces between hercynite and remaining staurolite. The gedrite intergrowths hence have a crystallographic orientation that is determined by hercynite. Possible reasons for this comprise the requirement of an incubation period before a critical nucleus can form and that nucleation only occurs when spinel-reactant silicate interfaces have formed to provide high-energy nucleation sites. The presence of Na in gedrite implies that it grew in contact with melt as also indicated by euhedral growth extensions into glass. There are no textures showing intergrowths between gedrite and Al-silicate to specify the relative timing of their formation after hercynite, but Al-silicate crystallized from the melt as demonstrated by its euhedral habit and glass inclusions. Newly formed quartz is always linked with Al-silicate suggesting that it crystalized as a result of local Si-saturation caused by the formation of Al-silicate. Zoned hercynite and ferrogedrite, Al-silicate compositions intermediate between sillimanite and mullite, "impure" quartz and glass that pseudomorph staurolite characterize a quenched breakdown assemblage that can be schematically described as

$$St + O_2 \rightarrow Hc + Gd + Sil + Qtz + L \tag{1.32}$$

Despite evidence of metastable melting, the pseudomorph mineral assemblage could mimic that of a possible stable melt-absent staurolite breakdown reaction

$$St = 3Hc + 0.2Gd + 5.6Sil + 1.2Qtz + 1.8V \tag{1.33}$$

as represented by the analyzed mineral compositions. However, the stoichiometric coefficients in the reaction are clearly in disagreement with the modal composition of the staurolite pseudomorph. This discrepancy may be ascribed to the presence of a siliceous peraluminous melt from which Al-silicate and quartz crystallized. Staurolite breakdown continued together with partial melting of muscovite, and plagioclase and the melt composition of reaction (1.32) was modified by element diffusion from these phases. The high alkali content of glass within the porphyroblast (2.1—4.6 wt%) suggests K from muscovite, that also contains between 1.47 and 1.72 wt% Na_2O, with additional Na and small amounts of Ca (0.08—0.81 wt% CaO) from plagioclase ($\approx$ An20) dissolution in the presence of quartz via an intergranular film of melt. High FeO (3.0—4.3 wt%) and lower MgO (0.35—0.49 wt%) of glass in the porphyroblast are inferred to be largely derived from staurolite, as pleonaste spinel is a reaction product of muscovite melting. Because the melt contains K, Na, and Ca from muscovite and plagioclase melting, the staurolite breakdown reaction could also be described in terms of the K—Na—Ca—Fe—Mg—Al—Si—H—O system, e.g.,

$$St + Ms + Pl + O_2 = Hc + Gd + Als + Qtz + L \tag{1.34}$$

The textural evidence described above led to the elucidation that the staurolite breakdown assemblage in the Wehr xenoliths is an illustration of a time—temperature—transformation with the sequence hercynite $\rightarrow$ gedrite/Al-silicate $\rightarrow$ quartz in the presence of melt. The Wehr xenolith appears to be the only known natural example where orthoamphibole is a replacement phase of staurolite, although its formation has been confirmed experimentally. Richardson (1968) observed the presence of minor amounts of orthoamphibole (an aluminous member of the ferro-anthophyllite—gedrite series) in experimental runs at 650 and 700°C/0.5 and 0.45 GPa, respectively, to produce Alm + Sil and Crd + Sill from St + Qtz. He suggested that, although this may point to a stability field for an amphibole-bearing assemblage, it is most likely metastable. Cordierite and almandine were not products of staurolite breakdown in the Wehr xenolith. The absence of cordierite may reflect pressure conditions of c. 0.34 GPa, whereas ferrogedrite (instead of almandine) suggests a minimum temperature of around 665°C at this pressure based on the reaction St + Qtz = OAm + Sil. Partial melting of muscovite, sodic plagioclase and quartz in the Wehr xenolith indicates temperature conditions above 645°C from the melting reaction

$$Ms + Qtz + Ab_{ss} + V = Als + L \tag{1.35}$$

From these data, minimum T-P conditions for the pseudomorph assemblage after staurolite in schist wall rocks of the Wehr magma chamber must have been 665°C and 0.35 GPa, i.e. around 20°C above the pelitic wet solidus, 20°C below the St + Qtz stability curve of Richardson (1968), and coincident with that of Dutrow and Holdaway (1989). If the temperature reached that of the melting reaction

$$Ms + Qtz + V = Als + L \tag{1.36}$$

($\approx 705°C$ at 0.35 GPa), as seems likely, then staurolite reactions, and the OAm + Sil-forming reaction would have been significantly overstepped by c. 75, 95 and 40°C, respectively, and the pelitic wet solidus temperature exceeded by c. 55°C A C−O−H fluid phase with H:O = 2, would lower the temperatures of reactions (1.35) and (1.36) by c. 10°C (Grapes and Li, 2010)

Xenoliths of metamorphic country rocks are locally profuse within breccia sequences in gabbroic to troctolitic rocks of the Voisey's Bay Intrusion, Labrador, Canada. Thermochemical interaction with mafic magma formed restite assemblages in the xenoliths that consist of Ca-rich plagioclase, corundum, hercynite and minor magnetite. Hercynite was formed due to the breakdown of garnet and pyroxene that were initially present in the metamorphic rocks, as well as by the replacement of corundum, which was itself the result of feldspar degradation during xenolith−magma interaction. Hercynite that substituted pyroxene and garnet is granular to bulbous, whereas replaced corundum is acicular to skeletal. Three separate morphologies of hercynite were observed in the xenoliths by Mariga et al. (2006). The first type of hercynite is acicular and resembles the morphology of corundum. Acicular hercynite forms composite grains with corundum in light-cored and variegated xenoliths. The second type of hercynite is granular in nature and is concentrated near the rims of the xenoliths. This kind of texture is also found in granular pyroxene from the contact aureole. The third type of hercynite is bulbous and shows a texture comparable to that of garnet in the contact aureole of the Tasiuyak Gneiss. Bulbous hercynite is generally closely associated with magnetite and occurs chiefly in the center of the xenoliths, surrounded by granular and/or vermicular textured hercynite. Vermicular and bulbous hercynite outline a texture mimicking cordierite−orthopyroxene symplectites that have been found in the Tasiuyak Gneiss within the contact aureole of the Voisey's Bay Intrusion. The composite grains of corundum and hercynite are inferred as proof for pseudomorphic replacement of corundum by hercynite. The close association of bulbous hercynite and garnet in the contact aureole, plus the fact that bulbous hercynite in the xenoliths outlines the same textures as garnet, is taken as confirmation that bulbous hercynite is a pseudomorph of garnet. This explanation is supported by the fact that bulbous hercynite is typically surrounded by granular and/or vermicular textured hercynite, defining a texture similar to the symplectitic intergrowths in the Tasiuyak Gneiss which formed during the early stages of garnet breakdown in the contact aureole. The fact that granular hercynites show very similar grain shapes and textures as granular orthopyroxene in the protolith assemblage is understood as confirmation for pseudomorphous replacement of pyroxene by hercynite. Pseudomorphous hercynite, alongside relict protolith textures such as crenulation cleavage and gneissic banding that are displayed by the xenoliths in the Voisey's Bay Intrusion, suggest that volume was preserved during mineral replacement reactions. The development of pseudomorphic textures call for the rate of dissolution of the precursor mineral to be similar to that of the concurrent precipitation of the new mineral, thereby conserving volume.

Complex multiphase pseudomorphs after perovskite (Nb-, LREE-poor) from calcite carbonatite (Sebljavr complex, Kola Peninsula, Russia) and serpentine calcite kimberlite (Iron Mountain, Wyoming) were described by Mitchell and Chakhmouradian (1998). In the kimberlite, the main products of perovskite replacement are (in order of crystallization): kassite, anatase and titanite plus calcite, ilmenite, LREE-Ti oxide [lucasite-(Ce)]. In the carbonatite, perovskite is at first replaced by anatase plus calcite and, later, ilmenite and ancylite-(Ce). In both cases, the growth of calcite and Ti-bearing phases after perovskite involved first progressive leaching of Ca^{2+} from the structure followed by crystallization of ilmenite and LREE minerals in the ultimate stages, after the precipitation of groundmass calcite. The development of kassite and titanite in the pseudomorphs on kimberlite was controlled by a lower Ca leach rate and higher SiO_2 activity in this system, in comparison to the carbonatite. The resemblance between the two types of pseudomorphs results from the instability of Nb-LREE-poor perovskite in a CO_2-rich fluid at low temperatures. The pseudomorphs are very comparable in every respect to their modal composition and mineralogy. In both types of pseudomorphs, perovskite as a primary phase experienced deuteric alteration, which resulted in a zoned aggregate of calcite and Ti-bearing phases. In both types, Fe-Nb-bearing anatase, calcite and Mn-bearing ilmenite are among the main products of alteration. The pseudomorphs from both occurrences reflect progressive loss of Ca from perovskite during the initial stages, and increase in Fe^{2+} and Mn^{2+} activities at the late stages of the alteration process. To finish, in both cases, LREE minerals are the latest phases to form. The pseudomorphs from the Iron Mountain kimberlite show a more intricate assemblage of replacement minerals than those from the Sebljavr carbonatite. Kassite and titanite are key constituents of the pseudomorphs from Iron Mountain but are not detected in the samples from Sebljavr. Compared to the Sebljavr carbonatite, anatase and ilmenite from Iron Mountain kimberlite characteristically contain variable amounts of Ca. The LREE released from the early perovskite were gathered in ancylite-(Ce) in the carbonatite, and in lucasite-(Ce) in the kimberlite. In the Sebljavr carbonatite and Iron Mountain kimberlite, the pseudomorphs after perovskite occur in a carbonate-rich paragenesis, in which calcite is associated with hydrous minerals (phlogopite and serpentine). Note that in both rocks, perovskite belongs to an earlier assemblage. In the carbonatite perovskite is a xenocrystic phase assimilated from the wallrock clinopyroxenite; in the kimberlite, it represents the early-forming groundmass mineral. Consequently the deuteric alteration of perovskite reflects the instability of this mineral in a CO_2-rich fluid. Interaction with the fluid involved progressive leaching of Ca from perovskite and its successive replacement by kassite or anatase (or both)

$$2CaTiO_3 + CO_2 + H_2O^- \rightarrow CaTi_2O_4(OH)_2 + CaCO_3 \tag{1.37}$$

Perovskite → kassite + calcite

$$CaTiO_3 + CO_2 \rightarrow TiO_2 + CaCO_3 \tag{1.38}$$

Perovskite → anatase + calcite

$$CaTi_2O_4(OH)_2 + 2CO_2^- \rightarrow TiO_2 + CaCO_3 + H_2O \tag{1.39}$$

Kassite → anatase calcite

The presence of anatase as a stable TiO_2 polymorph in both kimberlite and carbonatite cannot be elucidated with certainty. At ambient temperatures rutile is more stable than anatase or brookite. Therefore in most cases, the secondary TiO_2 polymorph developed after perovskite is labeled as rutile. As determined experimentally minor substitutions of Ti by other cations significantly affect the stability fields of anatase and brookite. Kassite, an intermediate product of conversion of perovskite to anatase (reaction schemes 1.37 and 1.39), is common in the Iron Mountain pseudomorphs but does not occur at Sebljavr. The simplest reason may include different Ca^{2+} leaching rates during the early stages of the alteration process. Another controlling factor is $P(CO_2)$ in the fluid.

The greater rate of leaching is thus anticipated in carbonatites, which is supported by the mineralogical observations. Note that the original difference in composition of perovskite from kimberlitic and carbonatitic source rocks is very small, and is doubtful to have any effect on the Ca^{2+} leaching rate and alteration pattern. In the Iron Mountain samples, anatase and calcite are closely associated with titanite. The simultaneous formation of anatase with titanite indicates some increase in SiO_2 activity at this stage of the process.

$$CaTi_2O_4(OH)_2 + (SiO_2)aq^- \rightarrow 2TiO_2 + CaTiSiO_5 + H_2O \tag{1.40}$$

Kassite → anatase + titanite

The absence of titanite in the Sebljavr pseudomorphs indicates the overall low activity of SiO_2 in the carbonatite system. The Sebljavr carbonatite is deprived of silicate minerals, and the occurrence of phlogopite in the vein is limited to its contacts with the country-rock clinopyroxenite. The growth of an ilmenite rim in both types of pseudomorphs shows an increase in $a(Fe^{2+})$ and $a(Mn^{2+})$ at the last stages of perovskite alteration. The increase in Fe^{2+} and Mn^{2+} activities apparently ensued from rapid precipitation of the groundmass calcite and corresponding decrease in $a(Ca^{2+})$. Note that in the kimberlite, the groundmass calcite was complemented by serpentine, whose formation used up most of the Mg present in the system. The crystallization of ilmenite after anatase can be described by the next reaction scheme:

$$TiO_2 + 2Fe^{2+} + 2H_2O \rightarrow 2FeTiO_3 + 4H^+ \tag{1.41}$$

Anatase → ilmenite

Perovskite is a major host for LREE in undersaturated ultramafic and alkaline rocks. In contrast to apatite, titanite and other possible carriers of the LREE in these rocks, perovskite exhibits a complete solid−solution series with the LREE-dominant counterpart, loparite-(Ce) (Chakhmouradian and Mitchell, 1997, 1998). Throughout the replacement of primary perovskite, the LREE are released into the fluid and form carbonate complexes stable in neutral and alkaline conditions. The precipitation of calcite and replacement of anatase by ilmenite (scheme 1.41) would lower the stability of these carbonate complexes, and initiate formation of the LREE minerals ancylite-(Ce) and lucasite-(Ce). As indicated above, these minerals are mostly restricted to the ilmenite-rich zone of the pseudomorphs. The occurrence of lucasite-(Ce), not carbonate minerals, as a major host for LREE in the Iron Mountain kimberlite apparently points to the co-occurrence of LREE-carbonate and Ti-bearing complexes in the late portions of the fluid (Mitchell and Chakhmouradian, 1998).

Margarite- and muscovite-bearing pseudomorphs after topaz from the Juurakko pegmatite dike, Orivesi, southern Finland, were described by Lahti (1988). A supercritical vapor phase rich in Ca and alkalis triggered alteration of topaz and some other silicates during the final phase of crystallization of the dike. The original columnar form of the topaz crystals is characteristic in the pseudomorphs, although roundish or irregular mica aggregates are also

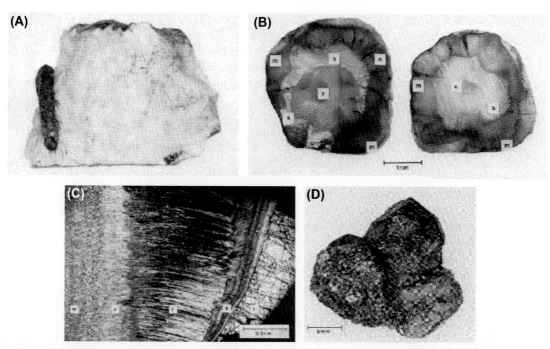

FIGURE 1.47 (A) Columnar mica pseudomorph after topaz crystal. Pseudomorph is homogeneous and consists only of very fine-scaled muscovite. (B) Two polished sections of the margarite-muscovite pseudomorph. The core is composed of a fine-scaled margarite-muscovite mixture (x) surrounded by a white irregular zone of fine-scaled margarite (s) and fine-scaled muscovite (m) or fibrous muscovite (n). The black radiating stripes are graphite. (C) A detail of the topaz-mica aggregate. The topaz crystal (t) is surrounded by the zones of fine-scaled margarite (s), fibrous margarite (f), fine-scaled margarite-muscovite (x) and fine-scaled muscovite (m). The fine-scaled margarite zone against topaz has several subzones differing in color and grain size. Microphotograph of a thin section, crossed nicols. (D) Two altered garnet crystals from the Juurakko pegmatite. The pseudomorphs contain only fine-scaled phengitic muscovite (Lahti, 1988).

common. The pseudomorphs vary extensively in size being 1–5 cm in diameter and up to 15 cm long. Sometimes they are closely associated with mica aggregates. The long prismatic form and the cross-section of the pseudomorphs similarly bare a resemblance to the form of andalusite crystals. Even though andalusite may occur in pegmatites, the mineral has not been found in the pegmatite dykes of the area, and the pseudomorphs are thought to be those after long prismatic topaz crystals. The pseudomorphs consist of fine-scaled, light-brown muscovite, but they can have a topaz-margarite or margarite core (Fig. 1.47). Coarse-scaled pink, lilac or yellow muscovite forms a rim around the pseudomorphs. The muscovite is nearly ideal dioctahedral. The muscovite zone around the topaz-margarite core consists of an inner, massive, fine-scaled subzone and an outer coarser subzone containing a pink lepidolite-like muscovite. Several columnar mica pseudomorphs with either poorly or well-developed crystal form approximating that of topaz were found in the pegmatite. The cross-section of the pseudomorphs is a square or oblique square. The prism faces are typically [110] faces of the altered topaz crystals, although combinations of [110] and [120] faces can be observed in some pseudomorphs. The pyramidal faces are generally deformed. The pseudomorphs may be homogeneous or zoned like the mica aggregates. The homogeneous pseudomorphs are composed of very fine-scaled massive muscovite, whereas the zoned pseudomorphs have a margarite-bearing core and may have some concentrically alternating margarite and muscovite zones; remnants of topaz were not observed. Pink, lilac or rose muscovites, which appear to be common in the pseudomorphs after topaz, are mostly very poor in iron, but the micas are enriched in manganese. The breakdown reaction of topaz to margarite can be written as:

$$\text{topaz} + \text{anorthite} + 2H_2O \rightarrow \text{margarite} + \text{quartz} + 2HF \tag{1.42}$$

The alternating zones of margarite and muscovite could be a consequence of diffusion during alteration. The occurrence of discontinuous shells of fibrous margarite around the topaz core could be explained by the mineral crystallizing in the crack or in a solution cavity opened between the topaz core and the surrounding muscovite shell. The fibrous form is typical of minerals crystallized during the opening of fractures of rocks. The pink muscovite surrounding the pseudomorphs was also created during the alteration processes of topaz. The mica is lesser in Fe and higher in Mn compared to the massive muscovite. The pink muscovites are somewhat enriched in Rb, Li, and Cs, most likely because of being the last micas of the pseudomorphs to crystallize. Of the trace elements, Rb, Cs and Ga have become mostly enriched in the muscovites, besides Li, Be, and Sr in the margarite. The pseudomorphs might also comprise iron-rich muscovite or hydromuscovite. Fine-scaled muscovite is also the principal mineral in the pseudomorphs after schorl and garnet. The pseudomorphs after topaz and tourmaline may be similar. The prismatic form and the hexagonal cross-section is, however, frequently well-preserved in the pseudomorphs after tourmaline and the muscovite has more Fe, Mg, Mn, and Ti. The tourmaline (schorl) crystals found in the intermediate zone, albite-rich fracture fillings and replacement bodies are frequently partially replaced by fine-scaled muscovite. Totally altered crystals are, however, sporadic, and the mica pseudomorphs regularly have a schorl core or small corroded inclusions of schorl. Corroded fragments of black tourmaline are common in one type of the pseudomorphs. The cross-sections of these pseudomorphs are roundish or partly triangular. The prismatic faces are uneven and the terminations irregular. The muscovite in the pseudomorphs after garnet has a considerable phengite component. Almandine—spessartine garnet has been encountered as an accessory mineral in several parts of the pegmatite. The crystals are euhedral or subhedral, ordinarily showing [211] faces. The mineral is generally fresh, but some muscovite pseudomorphs after garnet have been found in the albite—quartz pegmatite associated with the pseudomorphs after beryl, topaz, and tourmaline. The muscovite is fine scaled, with flakes 0.05—2.0 mm long and a brown yellow color. The pseudomorphs do not have fragments of garnet, but the original crystal form (icositetrahedron) of garnet is often well preserved. The pseudomorphs after beryl differ from the pseudomorphs after topaz and tourmaline in mineralogy, as they are composed of bertrandite or fine-grained bertrandite, chlorite and muscovite mass. The beryl found in the intermediate zone is fresh but in fracture fillings and albite-rich replacement units it is frequently altered. The size of the pseudomorphs varies extensively. Some of the beryl crystals are massive, up to 10—30 cm in diameter, and partially altered. Still, the smaller crystals may be entirely altered and filled with replacement and alteration products. Platy, colorless bertrandite is often the foremost mineral in the pseudomorphs (see Lahti, 1981), but porous pseudomorphs composed of massive chlorite—bertrandite—muscovite intergrowth are common too. The color of the mica intergrowths is greyish or brownish when stained with iron hydroxides. Comparable pseudomorphs after beryl have been described by Roering and Heckroodt (1964) from the Dernburg pegmatite, Karibib, South West Africa.

Bityite was observed by Lahti and Saikkonen (1985) in Li pegmatite dikes in the Eräjärvi area, southern Finland. The mineral is reasonably common in pseudomorphs after beryl or in cavities with bertrandite, fluorite, and FAP in the Maantienvarsi dyke. The uncommon minerals, including beryl, bityite, spodumene, cassiterite, Fe-tantalite, lepidolite, zircon, lithiophilite, and its alteration products, are enriched in the albite-rich pegmatite parts among the big microcline crystals or in small sugar albite or cleavelandite veins cross-cutting them. It is not easy to identify and can easily be confused with other micas. About three kilometers north of the Maantienvarsi pegmatite there is the renowned Viitaniemi pegmatite (Lahti, 1981), in which fine-grained greenish yellow muscovite (gilbertite) accompanied by beryllium phosphates (väyrynenite and hydroxyl-herderite) in pseudomorphs after beryl is a common mineral. The pseudomorphs filled with bityite are always small, 0.5—3.0 cm in diameter, and also comprise variable amounts of small, purple fluorite crystals and occasionally thin transparent bertrandite crystal plates, light green FAP, quartz and corroded fragments of beryl. Bityite has crystallized as an alteration product of beryl from fluorine-rich, probably acid hydrothermal solutions or fluids, although Černý (1968) has put forward that alumoberyllosilicates of alkalis and alkali earths are an indication for alkaline parent solutions.

The crystallization sequence and metasomatic alteration of spodumene ($LiAlSi_2O_6$), montebrasite ($LiAlPO_4(OH,F)$), and lithiophilite ($Li(Mn,Fe)PO_4$) were described for nine zoned lithium pegmatites in the White Picacho district, AZ, USA, by London and Burt (1982). Fracture-controlled pseudomorphic alteration of the primary lithium minerals is extensive and seemingly the product of subsolidus reactions with residual pegmatitic fluids. Spodumene has been changed to eucryptite, albite, and micas. Alteration products of montebrasite comprise low-fluorine secondary montebrasite, crandallite (tentative), hydroxyl-apatite, muscovite, brazilianite, augelite (tentative), scorzalite, kulanite, wyllieite, and carbonate-apatite. Secondary phases identified in altered lithiophilite include hureaulite, triploidite, eosphorite, robertsite, fillowite, wyllieite, dickinsonite, fairfieldite, Mn-chlor-apatite, and rhodochrosite. Original subsolidus metasomatism of the lithium minerals happened in an alkaline environment, as demonstrated by the albitization of spodumene and calcium metasomatism of the phosphates. The crystallization of secondary micas in spodumene, montebrasite, changed from alkaline to moderately acidic postmagmatic fluids, as (K + H)-metasomatism produced greisen-like or sericitic alteration. At the Independence and Midnight Owl pegmatites, abundant spodumene in contact with primary quartz shows a thin reaction rim of fine-grained albite, whereas the cores of the spodumene crystals are replaced to varying grades by a very fine-grained, fibrous intergrowth of eucryptite + albite. This original replacement formed bilaterally symmetrical veinlets of eucryptite + albite with their long fiber axes oriented at right angles to the {110} cleavage of the host spodumene. In some samples, eucryptite near veinlet centers and spodumene was then transformed to muscovite or lepidolite, causing the intergrowth of albite + mica called "cymatolite" by Brush and Dana (1880). Finally, remnant spodumene and secondary albite was transformed to lithian muscovite or lepidolite, producing soft, waxy pseudomorphs of almost pure mica after spodumene (referred to as "killinite" by Julien, 1879). The alteration series for spodumene from the White Picacho pegmatites and from other localities suggests that initial replacement involved Na^+-for-Li^+ exchange (as demonstrated by the crystallization of albite), followed by $K^+ + H^+$ metasomatism (bringing about the conversion of eucryptite, then spodumene and albite, to mica). Albitization of spodumene can be described by two reactions:

$$LiAlSi_2O_6 + SiO_2 + Na^+ = NaAlSi_3O_8 + Li^+$$

$$Spodumene + quartz = albite \tag{1.43}$$

which elucidates the thin reaction rims of pure albite separating altered spodumene from quartz, and

$$2LiAlSi_2O_6 + Na^+ = LiAlSiO_4 + NaAlSi_3O_8 + Li^+$$

$$Spodumene = eucryptite + albite \tag{1.44}$$

which explains the formation of eucryptite + albite intergrowths inside spodumene crystals. Whether spodumene breaks down to albite or to eucryptite + albite depends predominantly on the presence or absence of quartz, and therefore on the activity of silica inside and around large spodumene crystals The alteration of eucryptite to muscovite is assisted by the fact that the Al:Si ratios are identical in both minerals, and thus the development of pseudomorphic mica in spodumene is enhanced by a local decrease of silica activity. The direct transformation of spodumene to muscovite releases silica:

$$3LiAlSi_2O_6 + K^+ + 2H^+ = KAl_3Si_3O_{10}(OH)_2 + 3SiO_2 + 3Li^+$$

$$Spodumene = muscovite + quartz \tag{1.45}$$

At all of the pegmatites, montebrasite has been substituted by a number of secondary phases. Initial alteration of primary montebrasite involved hydroxyl exchange for fluorine, generating low-fluorine, secondary montebrasite along fractures and cleavages in the host primary phase. Similar to the Tanco pegmatite, Manitoba, replacement by secondary low-fluorine montebrasite is widespread in fresh looking material, but the primary and secondary phases are practically impossible to distinguish in hand specimen. Successive alteration of primary and secondary montebrasite formed fine-grained hydroxyl-apatite, muscovite, and small amounts of a calcium aluminum phosphate that appears to be crandallite (Fig. 1.48). Although these three secondary minerals occur in grain contact with montebrasite, the crandallite is consistently found at montebrasite borders and possibly was the first mineral to form. In a representative specimen, a thin reaction rim of crandallite at montebrasite borders grades outward into veinlets of greyish-white hydroxyl-apatite, subsequently an intergrowth of apatite + muscovite, and lastly fine-grained green muscovite in the middle of replacement veinlets and near the crystal boundaries of the primary phase. Several combinations of scorzalite, brazilianite, augelite (tentative), wyllieite, and kulanite are found as scattered crystals in the apatite + muscovite replacement that obviously cut the apatite + muscovite assemblage. Hence these rare phosphates formed late in the alteration sequence of montebrasite. A few colorless grains of an aluminum phosphate that could be augelite were observed in two specimens. The formation of secondary montibrasite by OH-for-F exchange in the primary phase probably reveals an increase in the ratio a_{H_2O}/a_{HF} in the residual fluid. Increase in this activity ratio implies depletion in fluorine (possibly as a result of crystallization of apatite, micas, or of the montebrasite itself) or a relative increase in a_{H_2O} with decreasing temperature. The early cation metasomatism of primary and secondary montebrasite involved Ca-for-Li exchange, generating hydroxyl-apatite and minor crandallite. Subsequent alteration to muscovite was far more extensive and reveals a high mobility of Li, K, Si, and P, but not Al. Despite their rarity in these pegmatites, the existence of montebrasite of scorzalite, kulanite, wyllieite, augelite, and brazilianite offers better understanding of the chemistry of the postmagmatic fluids. The scarcity and volumetric

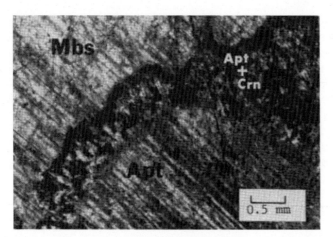

FIGURE 1.48 Photomicrograph (top) showing the replacement of twinned primary montebrasite (Mbs) by a fine-grained aggregate of hydroxylapatite (Apt) with minor crandallite (Crn). Notice in the lower right corner that the twinning present in the host phase has been preserved in the pseudomorphic replacement. Specimen is from the Midnight Owl pegmatite. Crossed polars (London and Burt, 1982).

insignificance of these three phosphates accentuates the extreme depletion in Fe (and enrichment in Mn) observed in these pegmatites. Of the five rare secondary phosphates mentioned above, brazilianite is the most abundant. Its occurrence late in the montebrasite alteration sequence indicates that its stability is comparable to that of lacroixite, $NaAlPO_4(F,OH)$. Lacroixite is incompatible with lithium aluminosilicates in the presence of quartz. In the White Picacho pegmatites, the presence of spodumene with albite and quartz probably buffered the a_{Na^+}/a_{Li^+} ratio in the co-existing fluid at values adequately low that brazilianite, like lacroixite, was unstable during most of the pegmatite recrystallization. Because montebrasite is found near or with spodumene + quartz in these pegmatites, the reaction of montebrasite to brazilianite probably happened only after spodumene had been locally destroyed (i.e., reacted to albite and/or muscovite). Lithiophilite shows two different replacement sequences, contingent on whether the lithiophilite is embedded in quartz or in albite. The least-altered crystals are found in quartz. In this association, the first replacement involved hydration and removal of lithium to create intergrowths of hureaulite and triploidite. Often, calcium-rich hureaulite is associated with eosphorite and fairfieldite. Fairfieldite also forms thin veinlets in close association with dark reddish-brown robertsite. Lastly in this association, robertsite and all other earlier-formed replacement phases were transformed to fibrous Mn-rich chlor-apatite. Relict subhedral lithiophilite crystals in albite show crude prismatic [001] forms, but these crystals are almost completely replaced. Cores of hureaulite + triploidite and minor lithiophilite are veined and replaced by fillowite, dickinsonite, and wyllieite. Fillowite has been reported before only from the pegmatite at Branchville, Connecticut (Brush and Dana, 1879) and from the Buranga pegmatite, Rwanda (von Knorring, 1963); in the White Picacho district, it occurs abundantly as massive, fine-grained beige crystal aggregates anywhere lithiophilite nodules are embedded in albite. Olive-green dickinsonite and yellowish-brown wyllieite usually occur; both phases usually are intergrown with fillowite. Fairfieldite generally forms thin veinlets that unmistakably cut the fillowite + dickinsonite + wyllieite assemblage. As in the montebrasite alteration order, fibrous coliform apatite forms bands that surround lithiophilite nodules and veinlets that cut all other secondary assemblage. Along with Mn-chlor-apatite, rhodochrosite forms masses and veinlets of pink crystals that characterize the last products of metasomatic alteration of lithiophilite crystals. Spongy masses of Mn-oxides, purpurite, stewartite, and strengite also replace lithiophilite, but this assemblage of secondary phases is not found in any freshly mined lithiophilite from subsurface excavations. These oxidized minerals seem to be the result of weathering instead of postmagmatic pegmatitic alteration. The early alteration of lithiophilite combined hydration with the removal of lithium to form hureaulite and triploidite. Subsequent to the formation of hureaulite and triploidite, alteration of lithiophilite nodules basically involved progressive Ca-for-Mn cation exchange. Dickinsonite, fillowite, and wyllieite are found only in lithiophilite nodules embedded in albite. Formation of these three phases reveals an increasing $a_{Ca^{2+}}/a_{Mn^{2+}}$ in the aqueous fluid, as well as localized high activity of Na^+. It seems that the occurrence of spodumene with albite and quartz kept values of a_{Na^+}/a_{Li^+} in the fluid adequately low that lithiophilite was not changed to natrophilite. By the time the ratio a_{Na^+}/a_{Li^+} had increased adequately to stimulate the growth of Na-rich phosphates (i.e., after spodumene had been destroyed by albitization), most lithiophilite probably had been altered to hureaulite + triploidite, and the activity of Ca^{2+} may have increased in the fluid. This condition seemingly led to the formation of dickinsonite, fillowite, and wyllieite from hureaulite and triploidite. Fillowite, dickinsonite and wyllieite are generally intimately intergrown, and this fact indicates that they formed contemporaneously. The ensuing crystallization of fairfieldite, robertsite, and apatite infers that $a_{Ca^{2+}}/a_{Mn^{2+}}$ increased during the alteration episode. Chlorine-rich apatites formed at this stage of alteration further indicate that the postmagmatic fluids in these pegmatites were F-deficient. The highly oxidized phases, such as purpurite, stewartite, strengite, etc., seem to be weathering products (London and Burt, 1982).

The tsumcorite-group mineral kaliochalcite, $KCu_2(SO_4)_2[(OH)(H_2O)]$, has been found in some fumaroles at the Second scoria cone of the Northern Breakthrough of the Great Tolbachik Fissure Eruption, Tolbachik volcano, Kamchatka, Russia. Two fumaroles, Yadovitaya and Arsenatnaya, are named as its type localities. The mineral appears to be the result of the interactions between high-temperature sublimating KCu-sulfates and atmospheric water vapor at temperatures not higher than 100–150°C. Kaliochalcite typically is found in polymineralic crusts (up to several dozens cm^2 in area and up to 0.5 cm thick), where it is usually the major component. It forms fine-grained pseudomorphs after anhydrous KCu-sulfates, mostly euchlorine, fedotovite, or piypite, commonly with their relics. Temperatures of 90–110°C were measured by Pekov et al. (2014) inside fumarole chambers, in the kaliochalcite-containing areas. In the inner regions of the same fumaroles, where temperature reached 270–430°C, merely anhydrous sulfates were found. Fine-grained kaliochalcite (with separate crystals usually less than 5 mm in size) is characteristically the major component of the crusts. These aggregates are frequently partial pseudomorphs after primary, anhydrous KCu-sulfates, mainly euchlorine $KNaCu_3O(SO_4)_3$, fedotovite $K_2Cu_3O(SO_4)_3$ or piypite $K_8Cu_9O_4(SO_4)_8Cl_2$ (Fig. 1.49), in some cases after kamchatkite $KCu_3O(SO_4)_2Cl$, alumoklyuchevskite $K_3Cu_3AlO_2(SO_4)_4$, or wulffite $K_3NaCu_4O_2(SO_4)_4$. This is clearly established by well-preserved crystal shapes and distinctive crystal clusters, besides the existence of abundant relics of these minerals in the pseudomorphs. In small voids of granular kaliochalcite aggregates. in the holotype specimen, kaliochalcite was found as light green complete pseudomorphs after groups (up to 2 mm across) of spear-like piypite crystals. Kaliochalcite forms only in the moderately hot, external sections of the fumaroles where it substitutes for hydrogen-free KCu-sulfates formed as sublimates under temperatures not lower than 360–380°C. Kaliochalcite contains OH groups and H_2O molecules and is associated with other hydrous sulfates and OH- and/or H_2O-bearing chlorides. This suggests that kaliochalcite was formed not by direct deposition from gaseous phase [gases in the fumaroles of the Second scoria cone of the Northern Breakthrough of the Great Tolbachik Fissure Eruption contain <1% water vapor (Zelenski et al., 2012)], but as the result of reactions between earlier, high-temperature sublimate sulfate minerals and atmospheric water vapor at moderately low temperatures, probably not higher than 100–150°C. The fumarolic gas might be a third component in this mineral-forming system that can be considered as a "mixed" one, i.e., combining features of fumarolic and supergene processes (Pekov et al., 2014).

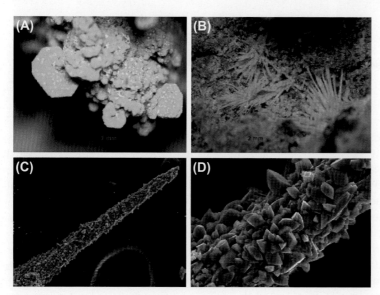

FIGURE 1.49   (A) Partial pseudomorphs of kaliochalcite after well-shaped fedotovite crystals. Yadovitaya fumarole. Field of view (FOV): 4.3 mm. (B) Complete pseudomorphs of kaliochalcite after bush-like groups of piypite crystals. The holotype specimen, Yadovitaya fumarole. FOV: 2.9 mm. (C) Complete pseudomorph of kaliochalcite after spear-shaped piypite crystal and (D) its magnified fragment. The holotype specimen, Yadovitaya fumarole. FOV: (C) 0.48 mm, (D) 0.12 mm. Scanning electron microscope (SEM) secondary electron images (Pekov et al., 2014 © www.schweizerbart.de). *(A and B) Photo: I.V. Pekov and A.V. Kasatkin.*

Potassium feldspar crystals from the Alto da Cabeca (Boqueirao), Rio Grande do Norte and the Morro Redondo, Minas Gerais pegmatites in Brazil have a typical adularia fourling habit consisting of a combination of Baveno and Manebach twins (Fig. 1.50). Crystal morphology suggests a monoclinic symmetry, but detailed XRD measurements and TEM show that the samples are actually low microcline. TEM and selected-area electron-diffraction investigations of the crystals show that they are compositionally homogeneous, without exsolution lamellae of other phases. The electron-diffraction patterns show no splitting of any reflection and no streaking, and thus indicate ideal order. Adularia pseudomorphs are highly ordered triclinic feldspars developed through transformation of a monoclinic parent. The term adularia is usually defined as a morphologically distinct variety of potassium feldspar that is typical of low-temperature hydrothermal environments. Most crystals of adularia exhibit intermediate degrees of order and structural state, whereas rare samples of adularia closely approach low microcline or high sanidine (Bermanec et al., 2012).

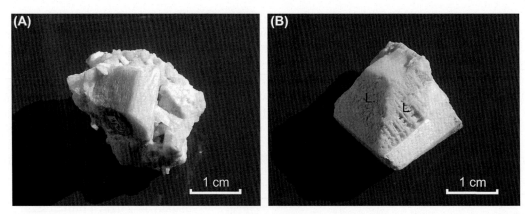

FIGURE 1.50   Potassium feldspar crystals from: (A) Alto da Cabeça (Boqueirão), RN and (B) Morro Redondo, MG; L: residual lamellae on the crystal face (Bermanec et al., 2012).

Clay pods from the Odd West (Bernic) and Buck Claim pegmatites in south-eastern Manitoba were identified as pseudomorphs after pollucite. The parent pegmatites belong to the pollucite-bearing group of the Bernic Lake–Rush Lake area, which gained world-wide recognition early last century for its size and its economic potential for lithium and beryllium. Braided micaceous veining preserved in the clay pseudomorphs closely resembles that of fresh pollucite from other localities. X-ray powder diffraction data corroborated by DTA and partial chemical analyses showed illite, kaolinite, quartz, and smectite minerals as the main components of the pseudomorphs. The association of illite + kaolinite + smectites in the clay pods leaves little doubt that it originated at low temperatures, very different from the conditions of crystallization of the enclosing intermediate blocky zone and core. The pseudomorphic character of the clay pods is best shown by the lepidolite veining, a feature that could not conceivably postdate the clay–mineral assemblage. A higher temperature precursor must have been penetrated by lepidolite prior to its breakdown to the clay assemblage. The alteration completely digested the primary mineral at the Buck Claim and Odd West localities but left the adjacent quartz, albite, K-feldspar, and amblygonite intact. Such a selective breakdown at low temperatures is characteristic of pollucite. The only other mineral known to succumb to this type of alteration is petalite; however, clay pseudomorphs after petalite display prominent parting inherited from the cleavage of the original mineral. The character of the alteration process is not clear, although hydrothermal action seems more likely than supergene agents (Černý, 1978).

Many ultramafic and ophiolitic rocks are hydrated to form partially or completely serpentinized rocks. The serpentinization is often particularly well developed in olivine-rich rocks such as dunites and some peridotites. Serpentinites are Mg-rich rocks resulting from metamorphic alteration of dunite, peridotite or pyroxenite and are composed essentially of serpentine minerals. Serpentine is the family name for three major minerals: lizardite (planar structure), chrysotile (cylindrical rolls) and antigorite (alternating wave structure). The serpentine minerals crystallize under different habits due to serpentinization processes: mesh and hour-glass pseudomorphic textures were formed from olivine, and thin-bladed pseudomorphic textures from pyroxene and amphibole crystals. Pseudomorphic textures have been described extensively in the literature (Caillaud et al., 2006 and references therein) as mesh, hour glass, and bastite assemblages (bastites are most commonly pseudomorphic textures after pyroxene and/or amphibole) composed mainly of lizardite, chrysotile and brucite (Caillaud et al., 2006). In some cases, transformation of olivine into serpentine is accompanied by a simultaneous transformation of orthopyroxene into bastite or talc and of clinopyroxene into actinolite, tremolite or other ouralitic minerals. Iddingsite is very variable in composition because it never consists of a single phase and represents a mixture of cryptocrystalline goethite (and possibly hematite) with a phyllosilicate (smectite, chlorite, talc (rarely), micas). Amorphous components may also be present. Alteration of olivine into iddingsite is complex and requires an addition of Fe^{3+} ions. Iddingsite is generally found as an exclusive alteration product pseudomorph after olivine in basic volcanic and sometimes in ultramafic coarse to medium grained plutonic rocks. It is never found as pseudomorphs after other primary minerals such as pyroxenes or amphiboles and never occurs as filling material in pores and cracks. Iddingsite may be found as pseudomorphs after olivine together with other secondary products such as serpentine or clayey minerals. Such transformations are the result of two or more successive alteration processes each of which effects part of the original olivine mineral (Delvigne et al., 1979). Microprobe analyses of serpentine pseudomorphs after olivine, orthopyroxene, and clinopyroxene have revealed partial inheritance of compositional characteristics from the parent phases. Bastite samples had higher Cr and Al contents than mesh-textured serpentine. Qualitative estimates of elemental mobility during such pseudomorphic serpentinization were Fe >> Al > Cr. The observed persistence of lizardite-bastite in progressively metamorphosed serpentinites suggests a compositional control. Serpentine pseudomorphs after olivine, orthopyroxene, and clinopyroxene were indistinguishable as groups on the basis of their Fe, Mg, Mn, and Ni contents. However, there were substantial dissimilarities with respect to Al and Cr, mainly in the comparison of mesh-textured serpentine with bastites. These observations seem to indicate that in pseudomorphic serpentinization, only the surfaces of the parent silicates participate in the reactions. If this premise is correct, it was argued that the chemistry of the pseudomorphic serpentines must be a function of partial or complete equilibration between serpentine and the aqueous fluid. Bastites after clinopyroxene and orthopyroxene have high Cr and Al compared to adjacent mesh pseudomorphs after olivine. This is interpreted as inheritance of the parent mineral composition due to the relative immobility of Cr and Al during low-temperature serpentinization. The distribution of iron (normalized to total octahedral cations) among adjacent serpentine pseudomorphs after olivine and pyroxene suggests a high mobility of Fe during serpentinization (Dungan, 1979).

Stevensite can be found at Dean Quarry, The Lizard, Cornwall, associated closely with the disordered, hydrated variety of talc known as kerolite. Both minerals occur as pseudomorphs after pectolite and as microcrystalline anhedral masses filling cavities and veins in gabbro. The Dean Quarry occurrence is similar to many others worldwide. Stevensite and kerolite formed after prehnite, calcite, analcime, pectolite, and natrolite during a phase of decomposition and alteration of these earlier minerals. TEM of the milled kerolite and stevensite—kerolite from Dean Quarry showed that the kerolite particles were mostly acicular reflecting the macroscopic morphology of the sample (pseudomorphous after pectolite). By contrast, the stevensite-rich sample was composed of small, thin, flaky particles. These observations are perhaps an indication of two mechanisms for the development of stevensite, namely dissolution—reprecipitation and dislocation-substitution for the transformation of stevensite from other minerals besides sepiolite (Güven and Carney, 1979). Alteration of pectolite to stevensite and kerolite occurred during a major phase of decomposition and dissolution that affected many preexisting minerals. Analcime in particular is often altered to a gray montmorillonite-type mineral. The occurrence of stevensite and kerolite in intimate association suggests that geochemical conditions were close to the boundary of the stability fields for these minerals. Various synthesis studies point to the formation of stevensite being favored by low Mg/Si ratios and alkaline conditions, whereas "hydrated-talc" is produced under acidic conditions at higher Mg/Si ratios (Güven and Carney, 1979). The presence of Na^+ stimulates the formation of stevensite, whereas the presence of Ca^{2+} promotes the formation of "hydrated-talc". One potential scenario might be that relatively acidic hydrothermal fluids leached Mg from the host gabbro and penetrated via existing calcite-pectolite veins. Initially kerolite or "hydrated-talc" precipitated following dissolution of the pectolite. With time, dissolution of calcite and precipitation of kerolite lowered the Mg/Si ratio and made the solution more alkaline thus favoring the formation of stevensite (Elton et al., 1997).

Stevensite aggregates, mostly octahedral in habit, occur in metamorphosed carbonatic ejecta of some pyroclastic beds on Procida Island, southern Italy (Fig. 1.51). Formation of the smectitic pseudomorphs is associated with the breakdown of periclase crystals as a result of late stage hydrothermal processes. The periclase together with forsterite and spinel, formed in an early stage of thermal metamorphism that resulted in calcite recrystallization and partial de-dolomitization in the carbonatic rocks. In regards to the occurrence of stevensite in the carbonatic xenoliths of Procida breccia, the mineralogical, petrographical and

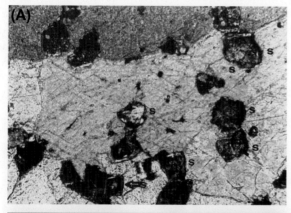

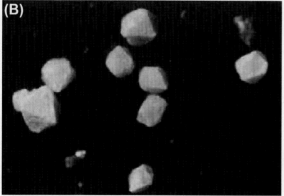

FIGURE 1.51 (A) Stevensite (S) aggregates in a carbonatic matrix; *arrow* points to an aggregate with perfect octahedral habit (polarizing microscope, crossed nicols, 65×). (B) Isolated stevensite pseudomorphs after periclase (binocular microscope, 65×) (Leoni and Sartori, 1983 © www.schweizerbart.de).

chemical data of the whole rock indicate a formation of the stevensite from periclase and, subordinately, from forsterite. The octahedral habit of the pseudomorphs and, especially, the comparison with the paragenesis of similar carbonatic ejecta from the neighboring Somma-Vesuvius area provides strong support for this interpretation. A crystallization of stevensite from the breakdown of spinel, with a similar octahedral habit, can be excluded both on the grounds of the always fresh look in thin section of this species and because of its chemical composition; which is characterized by high Al-content in sharp contrast with stevensite chemistry. The metamorphic events that influenced the carbonatic rocks of the Phlegraean Fields area ejecta could have operated in two successive stages. The first stage (of thermo-metamorphic character), characterized by high-temperature conditions ($>600°C$), involved, accompanied by recrystallization of calcite and a partial de-dolomitization, the crystallization of forsterite, spinel and periclase. The second stage, characterized by conditions of hydrothermal metamorphism (temperatures around $350°C$, pressures between 0.5 and 1.0 kbars, hydrothermal fluid circulation) brought on the breakdown of periclase (and partially of forsterite crystals) and crystallization of stevensite (Leoni and Sartori, 1983).

The Southern Illinois fluorite district has been an important domestic source of commercial fluorite for more than a century and is a prolific source of fluorite crystals and cleavages for collectors, schools and museums. Other minerals typical of Mississippi Valley type deposits such as calcite, galena and sphalerite are common throughout the district. The Minerva No. 1 mine, Hardin County, Illinois, located in the NW 1/4 of the SE 1/2 of Section 24, Township 11 S, Range 9 E, about 5 miles north of Cave-in-Rock, is typical of the mines in the Cave-in-Rock district which exploit fluorite ore bodies that have replaced limestone horizons. A special type of pseudomorph occurs as white hexagonal crystals to 2 mm in length associated with brown calcite. A study by Francis et al. (1997) irrefutably verified that they are replacement pseudomorphs of strontian baryte after paralstonite (Fig. 1.52). Originally there was speculation that the unknown mineral might be an intermediate member of the baryte-strontianite solid solution series. The XRD data were indexed and the unit cell refined, indicating baryte with strontium substituting for about one-third of the barium. The contradiction of this identification with the hexagonal-dipyramidal morphology of the crystals led to the identification of these crystals as pseudomorphs. Reexamination of the specimens under the binocular microscope showed that many of the crystals contain colorless, transparent cores. This material was identified as paralstonite by XRD. Thus the specimens were baryte pseudomorph after paralstonite. The Scanning Electron Microscope BSE image shows that the specimens consist of paralstonite that has been replaced by baryte. The texture reveals that replacement began at the surfaces of the paralstonite crystals and proceeded inward. Replacement of paralstonite by baryte is consistent with baryte being a late (probably the last) mineral to form in the Cave-in-

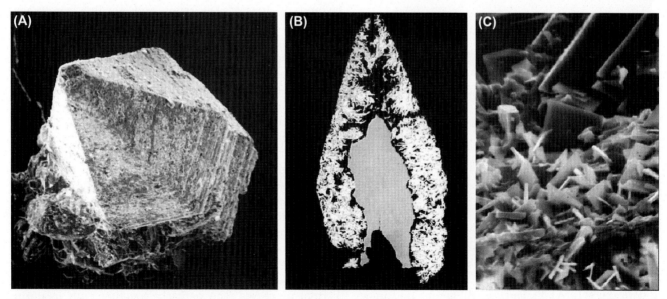

FIGURE 1.52   (A) Scanning electron microscope image showing the hexagonal dipyramid termination of a pseudomorph. Note the characteristic striations perpendicular to the c-axis on the "prism" faces. These striations can be used to distinguish paralstonite from calcite. (B) Back-scattered electron (BSE) image at 100× of a vertical section through a pseudomorph showing a homogeneous core (gray) of paralstonite partially replaced by bright white porous baryte. Specimen is 1.0 mm in length. (C) Scanning electron microscope image at 2000× of the surface of a pseudomorph which is a druse of tabular baryte crystals in fine and coarse sizes (Francis et al., 1997).

Rock paragenesis. Additionally, baryte pseudomorphs after witherite are known from the Minerva mine, and barite pseudomorphs after celestine were described by Lillie (1988) from the nearby Annabel Lee mine.

The pseudomorphs after the zeolites natrolite, chabazite, phillipsite, and gismondite were studied by Pöllmann and Keck (1993). These zeolites and pseudomorphs of them in the form of replacements by clay and mica minerals occur in cavities in a Tertiary basalt of Grosser Teichelberg in Upper Palatina, northern Bavaria in Germany. The following reactions were identified.

1. Replacement gismondite → clinochlor (leuchtenbergite) + calcite
2. Replacement chabazite → "Ca-illite" ± montmorillonite + calcite
3. Replacement phillipsite → "Ca-illite" ± montmorillonite + calcite
4. Replacement gismondite → montmorillonite + calcite
5. Cavity incrustation of phillipsite on gismondite
6. Cavity incrustation of montmorillonite on phillipsite
7. Incrustation pseudomorphs of montmorillonite on gismondite
8. Incrustation pseudomorph of montmorillonite on natrolite
9. Incrustation pseudomorph of phillipsite on gismondite
10. Pseudomorph of phillipsite after incrustation pseudomorph of montmorillonite to gismondite

The pseudomorphs of zeolites mainly consist of layer silicates. The chemistry of the new formed minerals exhibits an enrichment in MgO and some depletion in Al_2O_3 and CaO. Magnesium-ions are primarily enriched from the weathering effects of the surrounding basaltic materials. CaO is mostly fixed in secondary calcite. In comparison to other constituents no distinctive trends are apparent because the paragenesis of different minerals overlap. The zeolites are replaced step by step by various newly crystallizing minerals. The more stable Ca-poor zeolites form mainly incrustation pseudomorphs. Only in a few cases the complete zeolithic material was dissolved. Ca-rich zeolites primarily form replacement pseudomorphs. Previously described oriented intergrowth of phillipsite on gismondite crystals resulted in the cases of pseudomorphs in the formation of incrustation of phillipsite on totally pseudomorphosed gismondite. In some instances these pseudomorphs were dissolved and merely a cavity incrustation pseudomorph of phillipsite after gismondite was stable. The variable hardness of the pseudomorphs is mainly due to strongly intergrown calcite. Consequently no swelling by additional intercalation water molecules can occur such as with the clay minerals between the basaltic columns in the quarry. The stability of zeolites against replacement appears to impose the pseudomorphic trend of Ca-rich zeolites. Hence gismondite and chabazite are replaced relatively easy whereas phillipsite and natrolite mostly form incrustation pseudomorphs (Fig. 1.53). Pseudomorphs after gismondite are very common and more or less common after chabazite. Pseudomorphs after phillipsite and natrolite are actually sporadic.

Tungsten mineralization in hydrothermal quartz veins from the Nyakabingo, Gifurwe, and Bugarama deposits in central Rwanda has been found as the iron-rich endmember of the "wolframite" (ferberite-hübnerite) solid solution series (ferberite) and in the particular form of reinite. The occurrence of reinite is only reported from a few localities worldwide, for example, from the Cu-W-bearing tourmaline breccia pipe of Ilkwang (Korea), Kimbosan and Sannotake deposits (Japan), Mount Misobo (DRC), Rutsiro area (Rwanda), and several tungsten deposits in the Kigezi district in southwest Uganda (e.g., Ruhizha, Nyamalilo, Kirwa), which belong to the same "tungsten belt" and show very comparable geological characteristics to the Rwandan deposits. The scheelite crystals were replaced

FIGURE 1.53 (A) Scanning electron microscopy (SEM) micrograph of pseudomorph of clinochlor after gismondite with overcrustation of montmorillonite and phillipsite. (B) SEM-micrograph of incrustation pseudomorph of montmorillonite after phillipsite (Pöllmann and Keck, 1993 © www.schweizerbart.de).

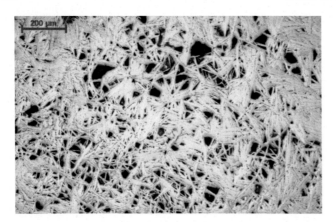

FIGURE 1.54 Photomicrographs in plane polarized reflected light. Reinite composed of randomly oriented fibrous ferberite crystals forming a porous aggregate (Goldmann et al., 2013).

by ferberite, forming pseudomorphs called reinite (Fig. 1.54). Reinite represents a variety of polycrystalline ferberite, which is usually present as euhedral aggregates still displaying the typical bipyramidal-tetragonal habit of scheelite. Because of the porous character of this type of ferberite, the bipyramidal aggregates are never intact, but the complete imprint of the bipyramid can still be observed in the surrounding quartz crystals. Microscopically, the ferberite aggregates are composed of individual small or even tiny fibrous crystals with dull-metallic luster. Remaining scheelite crystals inside the reinite were altered together with some of the ferberite crystals to different secondary tungstates, like ferritungstite, anthoinite, alumotungstite, cerotungstite, meymacite, raspite, and different Fe oxides. Ferberite crystallizes in the monoclinic crystal structure and commonly forms wedge shaped, bladed crystals, whereas scheelite has tetragonal symmetry and frequently occurs as bipyramidal crystals. Pseudomorphism of ferberite after scheelite is a common feature in Central Africa. Scheelite has been replaced by fibrous ferberite, a relationship that can be expressed through the following chemical reaction (1.46):

$$CaWO_4 + FeCl_2{}^{2-n} \leftrightarrow FeWO_4 + CaCl_m{}^{2-m} + (n-m)Cl^- \qquad (1.46)$$

This reaction could have been prompted by different mechanisms: the Ca/Fe ratio of the fluids in contact with the ore minerals, a change in the chloride concentration of the fluid, a lowering of the temperature and a change in pressure; although the pH value has a limited effect on the relative stabilities of scheelite and ferberite (Wood and Samson, 2000). During the stage of pseudomorphism of ferberite after scheelite, pyrite crystals in the quartz veins and the surrounding host rock have been altered leaving cubic cavities behind, now partly filled with Fe-oxides, such as goethite, limonite and secondary hematite. This process of pyrite alteration could be the origin of the iron for the replacement reaction (1.46) causing increased Fe concentration in the fluid and the lowering of the Ca/Fe ratio. The cubic cavities are now partly filled with Fe-oxides. Because the morphology of the scheelite crystals has been preserved, this replacement reaction occurred from the margin of the crystal to the core. The reaction preferentially started along crystallographic orientations of the earlier scheelite (Frisch, 1975), resulting in a boxwork network within the pseudomorphs. Because the molar volume of ferberite ($40.38 \text{ cm}^3/\text{mol}$) is smaller than that of scheelite ($47.05 \text{ cm}^3/\text{mol}$) (Robie et al., 1967), this replacement reaction resulted in the formation of pores, facilitating the continuation of the reaction. The primary ferberite is occasionally replaced by a second generation of skeletal and fibrous ferberite displaying abundant pores and cavities. The skeletal texture of this late ferberite may point to rapid growth from supersaturated and undercooled hydrothermal fluids. The primary ferberite crystals contain few pores, which only occasionally contain secondary tungsten minerals. On the contrary, the fibrous ferberite crystals are very porous, and these pores have in many instances been filled by secondary tungstates. The tungsten minerals, which are present inside these pores, are construed as alteration products. Two possible models have been put forward for the formation of these secondary tungstates. In a first model, the recrystallization reaction was incomplete, and as a result, some scheelite was preserved within the boxwork texture of reinite. Later, scheelite in these pores was altered to secondary tungstates. In a second model however, the reaction was complete, causing empty pores inside the reinite. In this model, the porous character and the small crystal size of the fibrous ferberite may account for the strong alteration compared to massive ferberite. In both models, alteration may have been caused by hydrothermal or supergene alteration processes. It is very well possible that both reactions have taken place simultaneously. Recent processes might have removed part of the alteration products, leaving some boxwork cavities in ferberite devoid of secondary tungstates (Goldmann et al., 2013).

Achtarandite, a pseudomorph after an unknown mineral of the point group of $\overline{4}3m$, was found as epitaxial overgrowth on zoned grossular crystals in the serpentinite from Yakutia, formed after skarned clay-carbonatic rock (Fig. 1.55). Hydrogrossular has been thought to have transformed by epitaxy to achtarandite. Achtarandite—a polycrystalline pseudomorph after an unknown tetrahedral mineral—has mystified mineralogists for more than 200 years (Galuskin et al., 1995). Different authors pointed out grossular, serpentine, saponite, kaolinite, chlorite, and carbonates as the main minerals in the pseudomorphs. Ca is an inert component in the skarn formation from lime stones.

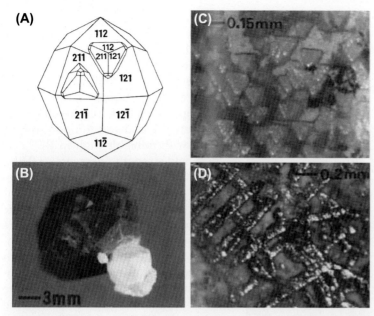

FIGURE 1.55 (A) Schematic drawing of epitaxy of achtarandite on grossular; (B) achtarandite [112] + [112] on grossular [211]; (C) epitaxial growth of hydrogrossular octahedra; (D) stitch-like, branchy epitaxial formations of hydrogrossular (Galuskin et al., 1995 © www.schweizerbart.de).

Intensive heating of the rock contributes to its dehydration and degasification (like in a cement process). High temperature favors the increase of Al diffusibility. According to Zaraysky (1989) the temperature must have been higher than 800°C. A sample chosen statistically from 200 green, yellow-green and brown-green grossular crystals with 542 grown white achtarandite crystals showed, that 95% of the achtarandite crystals had grown epitaxially on grossular. A prevalence of growth of achtarandite in the diffusion influence zone of grossular crystals (crystallization halo) demonstrates its coincidence with areas deficient in Si and rich in Al [for example: $12CaCO_3$ (calcite) $+ 3Al_4(Si_4O_{10})(OH)_8$ (kaolinite) $= 4Ca_3Al_2Si_3O_{12}$ (grossular) $+ 2Al_2O_3 + 12H_2O + 12CO_2$]. On the basis of the above mentioned facts Ca and Al are the main cations in the composition of protoachtarandite that hence can be classified as an oxide. Hypothetical composition (Ca, Al, O) and structure (close to grossular structure), well preserved morphology ($\bar{4}3m$) and paragenesis (grossular, diopside, vesuvianite), geological

position (contact with the thick diabasic dike) and hypothetical course of crystallogenesis and, finally, epitaxy on grossular (unit—cell parameter close to grossular—hydrogrossular) all point to the mineral mayenite ($12CaO \cdot 7Al_2O_3$, $I\bar{4}3d$, a = 11.98 Å). Most probably, wadalite ($Ca_6Al_5Si_2O_{16}Cl_3$) or chlormayenite ($\approx Ca_6Al_5Si_2O_{16.5}Cl_2$) was a protomineral for achtarandite (protoachtarandite) in which two atoms of Al were substituted by Si and charge balance was maintained by Cl. The subsequent zoning is typical for the internal structure of achtarandite pseudomorphs: (1) the outermost zone consists of a "chain-mail" layer, up to 0.3 mm thick, which is partially a pseudomorph and partially an epitaxial overgrowth; (2) the external zone is composed of pole-like hydrogarnet oriented subperpendicular to the surface of the achtarandite crystal. Thickness of the pole-like zone does not exceed a few millimeters in the 2—3 cm crystal. Each of the poles is formed by a series of {111} or {110} hydrogarnet crystals arranged as though they were threaded along a threefold axis. Space between the poles is mainly filled with serpentine; and (3) the internal part consists of irregular aggregates of minerals, mainly of the serpentine group. Hydrogarnet forms chains and curved plates, in which the subparallel orientation of the fourfold axes of crystal is preserved. Chlorite occurs as radial, concentric aggregates. Achtarandite can be classified as spongy (porous) pseudomorphs. Spongy pseudomorphs form in the system with isomorphic mixing of components; the "sponge" monocrystal is the end product in this case. The "sponge" is a monomineral polycrystal with the identical orientation of all crystals in the aggregate, formed during epitaxial substitution of wadalite ($Ca_6Al_5Si_3O_{16}Cl_3$, a = 12.001 Å) by hydrogrossular ($Ca_6Al_4Si_{12-x}H_4xO_{24}$, a = 11.962 − 12.061 Å) —minerals which are hydro- and chlorine-content phases of the metastable-isomorphic series grossular ($Ca_6Al_4^{VI}Si_6^{IV}O_{24}$)—mayenite ($Ca_6Al_4^{IV}[Al_3^{VI}\square_3]_6[O_{16.5}\square_{7.5}]_{24}$) (Galuskina et al., 1998). It is formed as a product of a "natural cement furnace" (thermal skarn). Hibschite is the major rock-forming mineral in rodingite-like rocks where it usually forms in achtarandite pseudomorphs within the serpentinite and chlorite layers. Morphological features and chemical composition of the hydrogarnet were related to the stages of development of the achtarandite pseudomorphs. The main evolution trend of the form and the composition of hydrogarnet were established: {111} hibschite → {110} Fe-hibschite → {211} Ti-hydrograndite. The {111} form of hibschite reflects the nonequilibrium conditions of crystallization at the early stage of achtarandite spongy pseudomorph formation. Its appearance was related to the stabilizing effect of a surface-active substance. The non-equilibrium nature of the octahedron controlled the features of its replacement by {110} faces together with the development of skeletal and translation forms. Replacement of {110} by {211} progressed gradually with the disappearance of rapid-growing {110} faces in the groundmass with an increase in the acidity of postmagmatic solutions; this

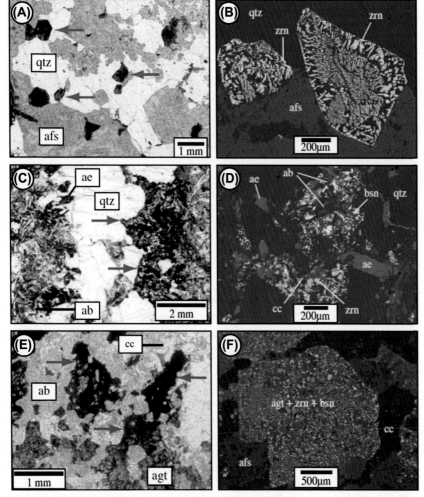

FIGURE 1.56 Photomicrographs of thin sections and scanning electron microscopy (SEM) back-scattered electron (BSE) images showing the three types of pseudomorphs. *Red arrows* point to the pseudomorphs. (A and B) Type-1 dark pseudomorphs disseminated in quartz (qtz) and alkali feldspars (afs). The SEM BSE image reveals the dendritic texture of zircon. (C and D) Type-2 pseudomorphs with quartz, aegirine and albite (ab), calcite (cc). The SEM BSE image shows small grains of bastnäsite-(Ce) (bsn) and zircon (zrn). (E and F) Type-3 pseudomorphs after aegirine (ae)—augite (agt) in the skarn. The replacing minerals are mostly bastnäsite-(Ce) and zircon (Estrade et al., 2014).

determined the increase of Fe content in hydrogarnet and changed the growth mechanism of its faces (Galuskina et al., 2001).

The Cenozoic Ambohimirahavavy alk. complex in the northwest of Madagascar consist, in its outer flanks, of a network of discontinuous granitic dikes with minerals high in REE and high-field-strength elements (HFSE). In coarse-grained granite (GR-I) prevails a magmatic miaskitic assemblage, rich in zircon. In the pegmatitic granite (GR-II), the rare metals are enclosed in secondary minerals, principally ameboid zircon and rare-earth fluorocarbonates, forming pseudomorphs after eudialyte. The eudialyte-bearing granite (GR-III), the richest in REE and HFSE, can be described by a magmatic agpaitic assemblage, dominated by eudialyte. A skarn formed at the contact of GR-I with limestone. Mineralization in this zone is in the form of pseudomorphs of rare-metal-bearing minerals after aegirine—augite. All rare-metal-bearing minerals recognized as secondary are miaskitic and were observed mostly in GR-II, where they are present as very small (10—50 μm) grains that nearly always form complete pseudomorphs after (a) precursor phase(s). These pseudomorphs are observed in sizes from less than 1 mm up to a few cm and usually have a well-defined crystal habit. Based on the nature of the replacing minerals, three pseudomorph varieties can be distinguished. Type-1 pseudomorphs have a distinctive tan color in hand samples, and generally occur in GR-II. Under the optical microscope, they look like euhedral to subhedral grains filled with a dark mass consisting of thin threadlike crystals. SEM BSE images show that these crystals are solely zircon, hosted in a quartz matrix and characterized by very unusual dendritic and botryoidal textures that exhibit a submicron-scale porous structure. Electronmicroprobe analyses of dendritic zircon reveal the presence of Al, Fe, and Ca. These elements are probably trapped in the pores as they are not compatible with the zircon structure (Fig. 1.56A,B). Type-2 pseudomorphs are the most common variety and are found mostly in GR-II, but are also locally present in GR-I. They range in shape from euhedral to anhedral with variable sizes and, in addition to zircon and quartz, may comprise several other rare-metal-bearing as well as nonrare-metal-bearing phases. Characteristic mineral associations include: zircon, unidentified Na-Ca zirconosilicate, fergusonite-(Y), REE fluorocarbonates [bastnäsite-(Ce), synchysite-(Ce), and parisite-(Ce)], monazite-(Ce), allanite-(Ce), cerite-(Ce), thorite, quartz, calcite, aegirine, albite, fluorite, and hematite. In these pseudomorphs, zircon is observed as both ameboid and subhedral grains. In some pseudomorphs, relics of the precursor mineral have been preserved, which in all cases were identified as eudialyte (Fig. 1.56C,D). Type-3 pseudomorphs were only found in GR-I that intrude a carbonate rock. Although type-3 pseudomorphs can be found through-out the entire thickness of the dyke, they mostly formed along its marginal parts, i.e., in the endoskarn, but were never observed in the exoskarn. Unlike the earlier described pseudomorph types, the precursor mineral is preserved in most cases, and could unmistakably be identified as a member of the aegirine–augite solid solution (Fig. 1.56E,F). The replacing minerals are made up of mostly extremely fine-grained ($\approx 10-20$ μm) aggregates of zircon and REE fluorocarbonates, plus minor amounts of fergusonite-(Y), allanite-(Ce), pyrochlore, monazite-(Ce), Nb-rich titanite and an unidentified zirconosilicate. These minerals are always found together with quartz, calcite, Fe–Mn carbonates, and, in some cases, also fluorite. In the exoskarn, very fine-grained REE fluorocarbonates are disseminated throughout an aggregate of diopside, andradite, calcite and wollastonite. In the agpaitic assemblage (GR-III), traces of alteration are limited to some grains of eudialyte that are partly replaced by the unidentified hydrated Na–Ca–zirconosilicate, depleted in REE and Nb with respect to the host mineral. In GR-II, effects of remobilization are ubiquitous in the form of pseudomorphs of rare-metal minerals after an early-formed precursor mineral. In type-1 pseudomorphs, the ameboid texture of zircon as well as its unusual composition (i.e., enrichment in Fe, Ca and Al) are direct evidence of the hydrothermal nature of the last (Geisler et al., 2007). On the other hand, type-2 pseudomorphs are characterized by a large diversity of minerals, including REE-and HFSE-bearing that in itself is diagnostic of a hydrothermal origin rather than magmatic reequilibration, which would have a tendency to produce a smaller number of phases. Because of this dissimilarity in texture and mineralogy, it is possible that type-1 and type-2 pseudomorphs originate from the alteration of different precursor minerals. However, the only relic mineral that was unequivocally identified as precursor was eudialyte in type-2 pseudomorphs. Experimental studies provide some understanding of the processes of alteration, for at least some REE and HFSE minerals, as well as the nature of the fluid transporting these elements. Geisler et al. (2007) indicated that, in the presence of a fluid phase, zircon reequilibrates by a diffusion reaction or by coupled dissolution–reprecipitation processes. The subsequent secondary zircon containing nano- and microscale porosity, where elements or minerals are trapped or precipitate, reminiscent of the textures observed in the zircon pseudomorphs. It is likely that an alkaline hydrothermal fluid altered the primary rare-metal minerals, particularly zircon and eudialyte. The constant presence of F in the different types of peralkaline granite, either as fluorite or as a major element of some primary minerals [nacareniobsite-(Ce), FAP] and secondary minerals (REE fluorocarbonates), indicates the abundance of this element, suggesting that the HFSE and REE were probably transported as fluoride complexes. However, it cannot be excluded that at least part of the REE were mobilized as chloride complexes. Furthermore, the silica saturation of the fluid, as shown by the fact that quartz is an abundant phase in the different types of pseudomorphs, has probably made this transport more effective. The most likely candidate for such an alkali-, silica-, Cl- and F-rich fluid would be a saline orthomagmatic fluid exsolved from the crystallizing peralkaline granitic melt (Estrade et al., 2014).

The Thor Lake rare element (Y–REE–Nb–Ta–Zr–Be) deposit, Northwest Territories, Canada, is a deposit hosted by alkaline syenite and granite and has two main mineralized zones, namely the Nechalacho and T Zones deposits. The T Zone was originally a pegmatite, but has experienced a strong hydrothermal overprint. Coarse-grained to megacrystic, elongate crystals outlining the pegmatitic texture, are fully pseudomorphed and have been interpreted to have initially been nepheline. The T Zone underwent three key alteration stages, which are (from earliest to latest): (1) silicification of the primary phases of the pegmatite and the associated replacement of igneous aegirine and arfvedsonite by quartz + magnetite + high field strength element (HFSE)-bearing minerals; (2) polylithionite alteration (Li metasomatism) that caused pervasive replacement of albite by polylithionite; and (3) phenakite alteration (Be metasomatism). Replacement of the primary mineralogy resulted in the creation of a number of pseudomorphs that are characterized by various HFSE minerals [typically, bastnäsite-(Ce), zircon, rutile, xenotime-(Y), and columbite] and phenakite (the main Be mineral) (Fig. 1.57). The habit and mineralogy of the pseudomorphs and the textural relationships in partially altered crystals, make aegirine and micas the most likely precursors of the HFSE-rich and Be-rich pseudomorphs, respectively. Although aegirine and arfvedsonite from alkaline rocks can contain HFSE, analysis of the aegirine and arfvedsonite from the T Zone showed that the HFSE concentrations in these minerals are too low to account for the bulk HFSE concentrations of the HFSE-rich pseudomorphs. Hence, HFSE must have been added to the pseudomorphs during fluid-mineral interaction, requiring hydrothermal mobility of HFSE. An open-space filling process may have been involved in the formation of HFSE minerals, including zircon, rutile, anatase, and REE fluorcarbonates. Initial replacement of aegirine by quartz + magnetite increased porosity in some of the T Zone rocks that may have aided HFSE precipitation. Although a precipitation model for HFSE involving fluorite (Salvi and Williams-Jones, 1990, 1996, 2006; Salvi et al., 2000; Williams-Jones et al., 2000) has been used for some deposits, here most hydrothermal zircons show no association with fluorite and the fluorite in REE-rich pseudomorphs mostly postdates the REE minerals. This shows that HFSE precipitation in the T Zone involves a model that does not rely on the mixing of HFSE-F-bearing fluids with Ca-rich fluids (Feng and Samson, 2015).

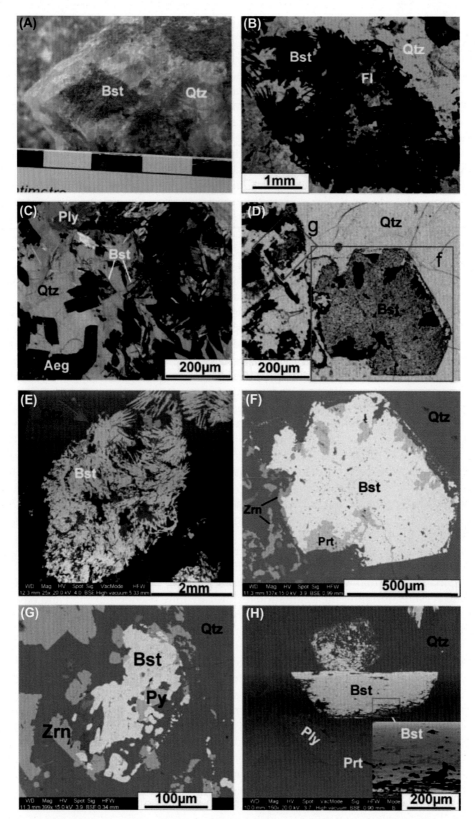

FIGURE 1.57   Images showing the mineralogy and morphology of bastnäsite-dominated pseudomorphs (Type 1). (A) Field photo showing bastnasite (Bst) aggregates in pseudomorphs with a rhombic habit in massive quartz (Qtz) (the length of the image approximately represents 5 cm). (B) and (E) Radiating bastnäsite that are rooted at the wall of the pseudomorph (*red arrows*) and have grown inwards. (C) Bastnäsite in a pseudomorph with a morphology that is identical to that of the altered aegirine (Aeg). (D, F, and G) Hexagonal to equant pseudomorphs in which parasite (Prt) has replaced bastnäsite; zircon (Zrn) rims the pseudomorphs. (H) A back-scattered electron (BSE) image showing that bastnäsite exhibits the same morphology as polylithionite (Ply) and that bastnäsite has partially been replaced by parisite. B and D are photomicrographs in plane-polarized light; (C) a photomicrograph in cross-polarized light; and (E–H) BSE images (Feng and Samson, 2015).

Pseudomorphs in Metamorphic Settings

Both magmatic and eclogitic parageneses are preserved in the gabbros of western Alpine ophiolites. The Allalin gabbro forms a 2 × 0.5-km outcrop above 3200 m between Zermatt and Saas Fee. It contains exceptionally preserved coarse-grained (up to 50 mm) magmatic textures in eclogitic rocks, and surprisingly kyanite occurs in talc-rich pseudomorphs after olivine. Four stages in the evolution of the Allalin gabbro can be recognized:

1. Rocks containing magmatic relics. High-P assemblages are confined to fractures or veins.
2. Coronites, with olivine or its pseudomorphs surrounded by garnet rims. Plagioclase in these rocks is pseudomorphed by jadeite-bearing assemblages.
3. Eclogite-stage rocks, in which olivine pseudomorphs comprise talc—kyanite or talc—Mg-chloritoid assemblages. Plagioclase sites now contain omphacite.
4. Posteclogite stage with the development of glaucophane, paragonite, and phengite in olivine sites.

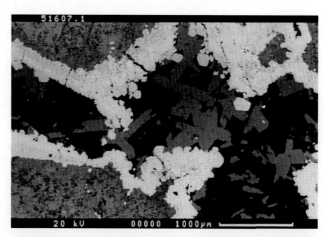

FIGURE 1.58 Back-scattered electron micrograph (SEM) of Allalin gabbro. Olivine site transformed to talc and omphacite. A rim of garnet separates the olivine site from pseudomorphed plagioclase (top and bottom left) (Barnicoat and Cartwright, 1997).

Undeformed gabbros with variable high-pressure overprints occur in a small area at the SW corner of the Rocciavre gabbro. A magmatic stage with olivine, in some cases rimmed by orthopyroxene, was followed by sea-floor alteration, forming brown hornblende after clinopyroxene and cummingtonite after orthopyroxene. A subsequent corona stage occurred, with plagioclase partially replaced by jadeite—zoisite—quartz, augite becoming rimmed by omphacite and garnet. Olivine may be partially pseudomorphed by talc, and is rimmed by garnet. Pseudomorphing of remaining magmatic minerals followed. Plagioclase is totally replaced by sodic pyroxene—zoisite—quartz, augite by omphacite and olivine by talc with garnet rims (Fig. 1.58). Gabbros occur in the Lago Superiore unit of the Monviso ophiolite where deformation is often mylonitic, but undeformed, eclogitized gabbros occur in low-strain augen. No magmatic relics have been reported, but garnet-bearing coronas are present around talc-rich olivine pseudomorphs. It has been suggested that water was introduced during high-pressure metamorphism. However, the similarity of olivine alteration patterns to those of ocean-floor gabbros suggests that hydration and local metasomatism leading to the stability of aluminous minerals in olivine sites occurred during hydrothermal alteration prior to subduction (Barnicoat and Cartwright, 1997).

A metagranodiorite from the ultrahigh pressure (UHP) Brossasco-Isasca Unit of the southern Dora Maira Massif, Western Alps, has preserved its original igneous microstructure owing to the absence of pervasive Alpine deformation and so metamorphic recrystallization was limited to the development of pseudomorphs and coronas at the expensive of the original mineral assemblage. During the UHP metamorphism the plagioclase was replaced pseudomorphically by jadeite + zoisite + kyanite + K-feldspar + coesite, whereas coronitic reactions occurred between biotite and adjacent minerals. The pseudomorph now consists of albite + jadeite + zoisite + kyanite + K-feldspar. Outside the original plagioclase crystals the peak coesite, which originally replaced igneous quartz, has been replaced by granoblastic polygonal quartz aggregates (Bruno et al., 2002). At the original igneous biotite-quartz contact a single corona of poorly zoned garnet was developed, whereas at the biotite-K-feldspar and biotite-plagioclase contacts, composite coronas were formed. The presence of zoisite among the plagioclase breakdown products required grain boundary diffusion of hydroxyl groups from nearby OH-bearing minerals, i.e., igneous biotite. The fact that quartz was not found as a plagioclase breakdown product indicates that Si was partly driven out of the plagioclase site by chemical potential gradients (Bruno et al., 2001).

The Les Essarts Unit (Armorican Massif, Vendee, western France) contains eclogites that are thought to be remnants of an old oceanic crust eclogitized during an Eo-Variscan subduction and later incorporated into the Variscan orogenic belt. They form kilometer-scale boudins within orthogneiss and paragneiss. The first episode was characterized by the intrusion of granite and migmatization of cordierite-bearing metapelites (T ≈ 670°C, P ≈ 0.32 GPa). The second episode is an eclogite-facies overprint (T ≈ 700°C, P > 1.6 GPa) that caused many pseudomorphic and coronitic reactions and resulted in high-pressure minerals to grow (garnet, phengite, kyanite, rutile, and probably jadeite) at the expense of the preceding high-temperature parageneses. The eclogite-facies pseudomorphs after cordierite, containing garnet + kyanite + quartz + micas, are rich in K$_2$O and deficient in MgO + FeO relative to a true cordierite (Fig. 1.59A,B); their compositions are similar to that of an altered cordierite (so-called "pinite"), indicating that cordierite had been altered to chlorite + micas before being replaced by the high-pressure metamorphic pseudomorphs. Pseudomorphs after cordierite: cordierite has been replaced by a cryptocrystalline aggregate of garnet (almandine−pyrope), quartz and kyanite, the presence of the latter having been confirmed by Raman spectroscopy. Some micas, biotite but also a minor white mica too fine-grained to be properly analyzed, have been also observed in very variable amounts. Pseudomorphs after sillimanite: The relict Al-silicate preserved inside the cordierite pseudomorph have now been replace by kyanite, but it was derived from sillimanite, of which it has kept the typical fibroblastic habit (i.e., "fibrolite"). Pseudomorphs after plagioclase: plagioclase forms a microcrystalline mosaic of poly-crystalline albite (Ab91−88), in which minute rods of kyanite, and minor apatite, zoisite and micas are only distinguishable with the SEM. The kyanite rods are ordinarily arranged in trails that delineate polygonal millimeter-sized cells with a honey-comb-like structure, which was interpreted as the outlines of the former plagioclase single crystals of the high-T paragenesis. The bulk composition of the pseudomorph, and so, ideally, that of the high-T plagioclase, was analyzed by scanning with the microprobe. It was found to be close to plagioclase in composition ($\approx$ Ab90−85: "plagioclase ps."), as was expected from the negligible modal abundance of the other phases in this pseudomorph. Garnet-bearing coronas at biotite−plagioclase interfaces (Fig. 1.59E): Garnet and phengite developed at the biotite−plagioclase interfaces, during a reaction that resulted in concentric coronas: [Biotite]/Phengite + Rutile inclusions/almandine-rich Garnet + Rutile inclusions/grossular-rich Garnet + Quartz inclusions/ [Plagioclase]. A distinct boundary occurs inside the garnet corona between two zones, (1) an inner almandine-rich rutile-bearing zone and (2) an outer grossular-rich quartz-bearing one. This boundary probably corresponds to the former biotite−plagioclase interface, as supported by the interruption of the grossular-rich zone where it encounters a primary quartz grain. In the polycrystalline plagioclase close to the corona, depletion in the Ca content confirms that anorthite$_{s.s.}$ was consumed to produce grossular in garnet. Garnet + rutile coronas at ilmenite− plagioclase interfaces (Fig. 1.59C,D): Ilmenite−plagioclase interfaces formed the following corona sequence: [Ilmenite]/Rutile/almandine-rich Garnet + Rutile inclusions/grossular-rich Garnet + Quartz inclusions/irregular Quartz corona/[Plagioclase]. Here again, an internal boundary in the garnet corona between two zones, with rutile

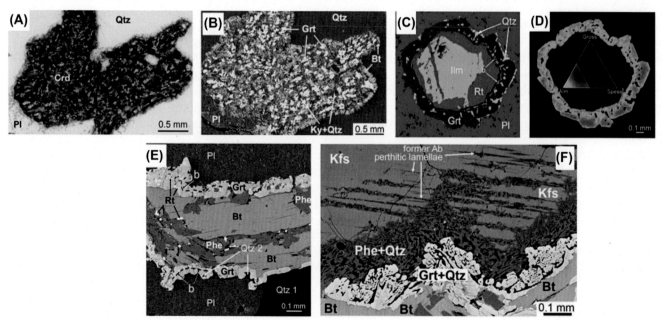

FIGURE 1.59 Pseudomorph after cordierite. (A) Plane-polarized light and (B) back-scattered electron (BSE) images. The pseudomorph (Crd) is made up of Grt + Qtz + Ky rodlets + micas flakes (mainly biotite, Bt). Corona between ilmenite and plagioclase. (C) Image produced by image analysis of X-ray element maps; line b: boundary between an inner rutile-bearing zone and an outer quartz-bearing one, interpreted as the original interface between ilmenite and plagioclase; (D) composition of the garnet corona obtained from calibrated X-ray maps of Ca, Mn and Fe (see Note 2 in SM), which shows the increase in the grs content toward the plagioclase side. (E) Corona between biotite and plagioclase. (A) BSE image; line b: boundary between an inner alm-rich rutile-bearing zone and an outer grs-rich quartz-bearing one, interpreted as the original interface between biotite and plagioclase. (F) Corona between K-feldspar and biotite. BSE image. The Phe + Qtz symplectite, grown preferentially along the planar perthitic exsolution lamellae, indicates that the albite exsolution (i.e., a first retrogression) occurred before the eclogite-facies metamorphism (Godard, 2009 © www.schweizerbart.de).

and quartz inclusions, respectively, most likely coincides with the former ilmenite—plagioclase interface, and a slight depletion in the Ca-content is once more detected in the polycrystalline plagioclase close to the corona. Coronas at contacts between biotite and microcline (Fig. 1.59F): double coronas grew at contacts between biotite and K-feldspar in the leucosome veins: [Biotite]/Garnet + Quartz/Phengite + Quartz/[K-feldspar]. The relative abundance of the Garnet + Quartz and Phengite + Quartz coronas is highly variable, as the Phengite + Quartz symplectite can almost miss or, alternatively, replace the whole K-feldspar. This could be due to the fact that the K-feldspar + $H_2O \rightarrow$ Phengite + Quartz alteration is not far from been isochemical, apart from for K_2O, a peculiarity that creates an almost singular matrix while balancing the whole reaction and prevents calculation of a stable solution. Apatite-bearing coronas around monazite only developed at contacts with plagioclase (Godard, 2009).

Metapelites from the southern aureole of the Vedrette di Ries tonalite (eastern Alps) were variably overprinted by contact and earlier regional metamorphic events during pre-Alpine and Alpine metamorphic cycles. In these rocks, starting from a primary garnet mica-schist (garnet stage), a complex series of transformations, affecting the site of the garnet, has been recognized. In the outermost part of the aureole, the primary garnet sites have been replaced by nodules of kyanite (kyanite stage). Closer to the tonalite, kyanite is replaced by staurolite (staurolite stage), which in turn is pseudomorphically replaced by muscovite (muscovite stage). Preservation of the subspherical garnet shape during all these transformations and persistence of mineralogical and textural relicts from earlier stages were favored by the very low strain experienced by the rocks because the garnet stage. Based on the appearance, disappearance and changes in abundance of minerals in nodules and matrix, model reactions can be proposed for some of the transitions between stages. Transition of the kyanite to the staurolite stage is of the type Kyanite + Garnet + Muscovite + Chlorite = Staurolite + Andalusite + Biotite, whereas the change from the staurolite to the muscovite stage is of the type Staurolite + Andalusite + Chlorite + Muscovite = Biotite + Sillimanite. Both reactions involve two Al_2SiO_5 polymorphs: given the polymetamorphic evolution of the rocks, and the distance between samples, they should not be considered as real equilibria in the AFM system (Cesare, 1999; Fig. 1.60).

A local equilibrium, irreversible thermodynamic model was applied by Foster (1983) to study biotite-rich pseudomorphs after staurolite in muscovite-bearing politic rocks near Rangeley, Maine. This kind of pseudomorph is

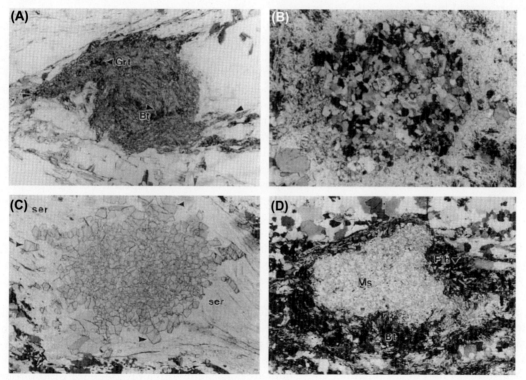

FIGURE 1.60 (A) Round aggregate of microgranular kyanite with minor biotite, ilmenite and garnet. Two kyanite-rich asymmetric tails (*arrows* extend in the subhorizontal main foliation. Plane-polarized light (PL), width of view (wv) = 6.5 mm. (B) Nodule of kyanite and staurolite. Kyanite occurs at the core of most staurolite crystals (asterisks). Crossed polars (CP), wv = 6.5 mm. (C) Monomineralic nodule of staurolite in a sericite felt (ser). Note (*arrows*) the coarser grain size of staurolite in the periphery of the aggregate, or isolated in the sericite matrix. Sericite around nodule (lower right) shows preferred orientation with incipient shear band cleavage. This suggests reactivation of the main foliation during contact metamorphism. PL, wv = 6.5 mm. (D) Decussate monomineralic aggregate of muscovite wrapped by folia of biotite and fibrolite. PL, wv = 6.5 mm (Cesare, 1999).

found near sillimanite segregations that have formed during a staurolite breakdown reaction. Biotite-rich, plagioclase-rich, muscovite-free pseudomorphs were formed in the model when the muscovite-free mantle of a sillimanite segregation surrounds a staurolite poikiloblast or grows over a muscovite-rich pseudomorph. The model predicted biotite-rich, muscovite-free, plagioclase-free pseudomorphs when the muscovite-free, plagioclase-free mantle of a sillimanite segregation enclosed a staurolite poikiloblast or grew over a muscovite-rich or plagioclase-rich pseudomorph. Biotite-rich pseudomorphs after staurolite in rocks from the lower sillimanite zone new Rangeley, Maine, are similar to those calculated by the model. The model predicted: (1) muscovite-rich pseudomorphs with small amounts of biotite, plagioclase and quartz form if a staurolite poikiloblast dissolves in a matrix where muscovite, biotite, plagioclase, and quartz are present (Foster, 1981) (Fig. 1.61); (2) muscovite-rich, plagioclase-free pseudomorphs containing small amounts of biotite and quartz form when the plagioclase-free mantle of a sillimanite segregation consumes a matrix pseudomorph or when a staurolite poikiloblast dissolves while surrounded by the plagioclase-free mantle of a sillimanite segregation;

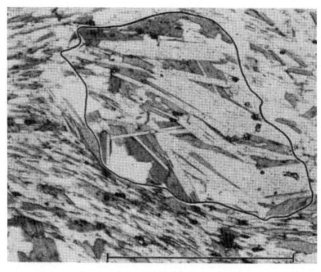

FIGURE 1.61 Muscovite-rich pseudomorph after staurolite (51% muscovite, 26% biotite, 14% plagioclase, 8% quartz, 1% ilmenite). The boundary of the pseudomorph is shown by a *solid line*. The small, high relief blebs in muscovite in the pseudomorph are remnants of staurolite. The scale bar is 1 mm (Foster, 1983).

(3) biotite-rich, plagioclase-rich, muscovite-free pseudomorphs containing quartz develop when a matrix pseudomorph is consumed by the muscovite-free mantle of a sillimanite segregation or when a staurolite poikiloblast dissolves while surrounded by the muscovite-free mantle of a sillimanite segregation; (4) biotite-rich, plagioclase-free, muscovite-free pseudomorphs containing quartz form when a pseudomorph is consumed by the muscovite-free, plagioclase-free mantle of a sillimanite segregation or when a staurolite poikiloblast dissolves in the muscovite-free, plagioclase-free mantle of a sillimanite segregation. According to the model, all four types of pseudomorphs may exist in one rock. Comparison of the modeled textures with those observed in specimens from the lower sillimanite zone near Rangeley, Maine showed that all four types of pseudomorph textures do occur. Pseudomorphs that are muscovite-free or plagioclase-free are rare in the Rangeley rocks because the mantles around the sillimanite segregations are thin, seldom enclosing staurolite pseudomorphs.

Pseudomorphous aggregates of muscovite after staurolite are characteristic of the lower sillimanite-to upper sillimanite-grade metapelites in parts of northwestern Maine, USA (Fig. 1.62). Traditional explanations for these pseudomorphs included retrograde metamorphism or K metasomatism. However, mineralogical, petrographical, and field evidence clearly indicates that the pseudomorphs have formed as a by-product of a simple prograde metamorphic event. The exact mechanism by which the pseudomorphs developed remains uncertain but seems to involve the dissolution of staurolite and concomitant replacement by muscovite. In the upper staurolite zone, moderately poikilitic, 1 cm subhedral to euhedral staurolite can be found in a medium-grained, well-foliated matrix of muscovite, biotite, quartz, and plagioclase. Coinciding roughly with the first appearance of sillimanite, staurolite in many specimens becomes anhedral with coarse laths of muscovite occurring around the outer rim. At gradually higher grades in the lower sillimanite zone the muscovite rimming becomes more prominent and the enclosed staurolite shrinks until it disappears at the isograd marking the upper sillimanite zone. The degree of pseudomorphing is readily recognizable even in the field. In some instances the pseudomorphs show shapes typical of staurolite twins and sometimes include several up to 2 mm garnets. Above the isograd marking the upper sillimanite zone the aggregates of muscovite in the pseudomorphs are likely to recrystallize into single large plates, up to 2 cm across, and commonly lie at high angles to the enclosing foliation. In some cases these muscovite plates include swarms of fibrolitic sillimanite. These data are consistent with a prograde transition from the staurolite to upper sillimanite zone. The prograde reactions defining the isograds can be given as: Upper Staurolite to Lower Sillimanite Zone

$$\text{staur} + \text{Mg-chlor} + \text{sodic-musc} \leftrightarrow \text{sill} + \text{bio} + \text{K-richer musc} + \text{Ab} + \text{qtz} + \text{H}_2\text{O} \tag{1.47}$$

Lower Sillimanite to Upper Sillimanite Zone

$$\text{staur} + \text{musc} + \text{qtz} \leftrightarrow \text{sill} + \text{bio} + \text{K-richer musc} + \text{Ab} + \text{H}_2\text{O} \pm \text{garn} \tag{1.48}$$

FIGURE 1.62 Early stages of muscovite pseudomorphism: (A) incipient muscovite plates forming around staurolite. (B) Coarse muscovite plates completely rimming remnant staurolite grains. Note sharp grain boundaries. (C) Later stage of muscovite pseudomorphism. Completed pseudomorph of muscovite flakes after stauolite. Fibrous material in upper left is sillimanite (Guidotti, 1968).

Such reactions suggest prograde metamorphism. No direct evidence supports an influx of K^+ as an explanation for the pseudomorphs. The development of muscovite pseudomorphs after staurolite probably includes nucleation phenomena. Pankiwskyj (1964) suggested that the dissolution of the staurolite could create a pressure void into which muscovite might readily recrystallize. It is obvious, according to reactions (1.47) and (1.48), that muscovite is involved in the reactions and thus in an activated state. Under such circumstances it might be expected to readily go into solution and then reprecipitate by replacing staurolite that is dissolving. The consequence will be a decrease in free energy inasmuch as less but larger muscovite grains form. Furthermore, the reaction is progressing with a rise in T, thus favoring a more K-rich muscovite. If the new muscovite forming in the pseudomorphs is even marginally more K-rich than the groundmass muscovite this will tend to make the groundmass muscovite somewhat metastable and so increase its likelihood of dissolution and then reprecipitation as a more K-rich muscovite in the pseudomorphs (Guidotti, 1968).

Optically homogeneous pseudomorphs can be found in the high-pressure metapelites from the Alpine Rif chain, Morocco, showing a complex metamorphic replacement sequence of a previous high-pressure metamorphic mineral, possibly magnesiochloritoid. Within the pseudomorphs, three zones were recognized based on the occurrence of distinct layer silicates, which crystallized under different Pressure—Temperature (P-T) conditions along the retrograde metamorphic P-T path. These layer silicates consisted of, from core to rim: a relic mixture of margarite and chlorite, oriented chlorite-smectite intergrowths ± pyrophyllite, and chlorite and abundant smectite (at the periphery of the pseudomorph). These pseudomorphs represent an uncommon record of the retrograde metamorphic P-T path because the earliest smectite formed while the rocks still experienced P-T conditions in the stability field of pyrophyllite. This clearly shows that clay minerals may crystallize as regional metamorphic products, and their stability field might not be restricted to late shallow-depth alteration (Agard et al., 1999).

Fiordland in New Zealand exposes the lower crustal root of an Early Cretaceous magmatic arc that now forms one of Earth's most extensive high-pressure granulite facies belts. The Arthur River Complex, a dioritic to gabbroic suite in northern Fiordland, is part of the root of the arc, and records an Early Cretaceous history of emplacement, tectonic burial, and high-pressure granulite facies metamorphism that accompanied partial melting of the crust. Late random intergrowths of kyanite, quartz and plagioclase partially pseudomorph after plagioclase and hornblende in the earlier high-temperature assemblages of the Arthur River Complex, indicate high-pressure cooling of an overthickened crustal root by ~200°C. The kyanite intergrowths are themselves partially pseudomorphically replaced by paragonite, regularly in the presence of phengitic white mica. The thermobarometric results point to the kyanite, paragonite and phengite-bearing assemblages evolving in the deep crust (P = 11—13 kbar) at temperatures of around 600—700°C. The paragonite and phengite-bearing assemblages suggests that water played a vital role in the development of the white mica-bearing assemblages. However, the available thermobarometric techniques give no information with respect to the proportion of fluid that accompanied the development of the kyanite and paragonite (with or without phengite) bearing assemblages. Biotite-plagioclase intergrowths that formed partially pseudomorphs after phengitic white mica and diopside-plagioclase intergrowths that partially pseudomorphosed jadeitic diopside, combined with published thermochronology results, are consistent with later rapid decompression (Daczko et al., 2002).

Prograde tourmaline-rich muscovite pseudomorphs after staurolite formed in sillimanite zone metapelites nearby to peraluminous granitoid intrusives in NW Maine, USA (Fig. 1.63). Tourmalines occur in discrete domains restricted to central regions of muscovite-rich, quartz-poor pseudomorphs with biotite-rich rims. A possible explanation concerns fluid-infiltration as the contributing mechanism for the formation of these tourmaline-rich mica pseudomorphs after staurolite. Irreversible thermodynamic models of local reactions and material transport in combination with mineral chemistry allow for an evaluation of reaction mechanisms that produced these pseudomorphs. Thermodynamic models in the Na-K-Ca-Mg-Ti-Fe-Al-Si-H-O-B system represented the observed textural features if a three-stage process was used. In stage 1, staurolite replacement was initiated by infiltration of an aqueous fluid that added $K + Na + H_2O$ to the rock with the concomitant removal of $Al + Fe$. Because the system was initially undersaturated with respect to tourmaline, a pseudomorph containing muscovite with minor biotite formed at the expense of staurolite. In stage 2, with continued infiltration, the concentration of B increased, tourmaline saturation was exceeded, and tourmaline nucleated and grew. Local material transport constraints mandate that tourmaline precipitation be spatially restricted to regions of staurolite dissolution. Consequently, tourmaline formed in clusters at sites containing the last remnants of staurolite in the pseudomorph core, also demonstrated by staurolite inclusions within several tourmaline grains. Resultant domains of staurolite replacement formed during this stage comprise about equal amounts of muscovite and tourmaline. Typical staurolite poikiloblast pseudomorphing reactions require silica transport. Matrix quartz dissolved from the surrounding host rock resulting in a local enrichment of biotite and plagioclase at the pseudomorph margin. In stage 3 small amounts of sillimanite nucleate and grow throughout the rock. Late-stage aqueous fluids from the adjacent monzonitic intrusive are likely to have been the primary B source. The above observations and analyses suggest that the pseudomorphs developed in a system open to an infiltrating fluid that served as a source/sink for components consumed/produced by the pseudomorphing reactions. The fluid appeared to provide K because sillimanite, which normally forms from the Al liberated by staurolite breakdown, is much lower in abundance than the amount estimated to develop in a closed system. Instead, the Al produced by staurolite break-down is consumed by micas formed as a result of K metasomatism. In addition, the infiltrating fluid was also assumed to be low in B initially but higher in B later on because the concentration of tourmaline in the pseudomorph center suggests that tourmaline growth began after staurolite had already been partially pseudomorphed by micas. The dissolving staurolites were assumed to link with the infiltrating fluid reservoir by diffusion because the existence of biotite + plagioclase-rich mantles around the pseudomorphs shows that material transport to local reaction sites was diffusion controlled. The sum of the local reactions that created the staurolite pseudomorph produced Fe, Mg, and Al while consuming Na, K, Ca, and Ti. These components were supplied or removed by diffusional transport through the matrix around the staurolite. In typical sillimanite zone rocks, components produced/consumed by the staurolite-pseudomorphing reactions were consumed/produced by local reactions growing sillimanite elsewhere in the rock, so that the overall reaction was balanced at the whole rock scale for all components except H_2O, which migrated out of the rock. Rocks with abundant tourmaline in the pseudomorphs typically do not have sufficient sillimanite to balance the components produced or consumed by the staurolite-pseudomorphing reactions, indicating they were open to infiltrating fluids with K (Dutrow et al., 1999). Seventeen sillimanite-zone

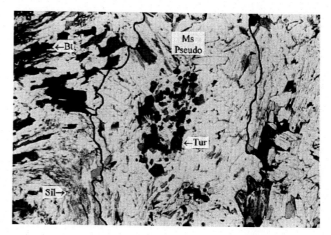

FIGURE 1.63 Photomicrograph of a tourmaline-bearing, muscovite-rich pseudomorph after staurolite that displays the change in grain size and mineral modes between matrix and pseudomorph; ms = light tabular grains within pseudomorph; bt = dark tabular grains; sil = gray needles. *Dark line* outlines pseudomorph. Note the clustering of tourmaline (tur), dark grains, within the center of muscovite-rich area (field of view = 5.2 mm) (Dutrow et al., 1999).

metapelitic samples, each containing two to five muscovite-rich pseudomorphs, from the Farmington Quadrangle, Maine, USA were investigated to evaluate their use as indicators of conditions during prograde metamorphism by Dutrow et al. (2008). Based on modal mineralogy determined from image analyses, the muscovite-rich pseudomorphs were allocated to four major types: (1) muscovite-rich (>70% muscovite), (2) quartz-muscovite (60%−70% muscovite with 10%−25% quartz), (3) plagioclase-muscovite (58%−72% muscovite with 10%−20% plagioclase), and (4) sillimanite-plagioclase-muscovite (50%−60% muscovite with 10%−20% each plagioclase and sillimanite). A biotite-rich, muscovite-poor layer surrounds many pseudomorphs. All pseudomorphs were interpreted to be prograde, based on their texture, and are after staurolite because of the partial replacement of staurolite by coarse muscovite at lower grades (Dutrow et al., 1999). Textural modeling by Dutrow et al. (2008)of the reaction mechanisms required to replicate the observed mineralogy in the pseudomorphs showed that each major pseudomorph type holds clues to the prograde path and signifies a different mechanism of formation. Muscovite-rich and quartz−muscovite pseudomorphs formed by the breakdown of staurolite that contained different modal amounts of poikiloblastic quartz. Quartz−muscovite pseudomorphs likely reflect a quartz-rich initial rock composition. Plagioclase−muscovite pseudomorphs call for the infiltration of sodium-bearing fluids. Sillimanite−plagioclase−muscovite pseudomorphs involve a two-stage process; the permeation of sodium-rich fluids during staurolite breakdown followed by sillimanite growth. The subtle mineralogical differences documented in the pseudomorphs provide evidence of previously unrecognized controls along the prograde path during metamorphism. Guidotti (1968) first recognized the significance of muscovite pseudomorphs after staurolite. Dutrow et al. (2008) established that careful analyses can extend his original studies to determine the details of the processes responsible for their occurrence. These data provided additional insights into the prograde conditions within the Farmington area of Maine that previously have gone unrecognized, and emphasized the utility of pseudomorphs and their subtle mineralogical variations in unraveling metamorphism.

Cordierite−phlogopite gneiss with coarse pseudomorphs of sillimanite after andalusite occurs as a kind of Mg-Al-rich rock in the Dabeeb−Hytkoras area of the granulite-faces western Namaqualand Metamorphic Complex, South Africa. The sillimanite pseudomorphs are all the time fringed by spinel−cordierite coronas, which, subsequently, are surrounded by cordierite monomineral zones. The spinel−cordierite coronas contain small amounts of sapphirine, corundum, and rutile. The coarse sillimanite pseudomorphs consist of several sectors that comprise of a mosaic of typically subparallel prisms of sillimanite. The prismatic habit of the earlier andalusite crystals is well preserved in the low-strain part. Alignment of (010) and [001] of sillimanite prisms is linked to the outlines of the former andalusite. The sillimanite mosaic seems to have formed by polynuclear recrystallization from andalusite. Small amounts of corundum, sapphirine, spinel, cordierite, phlogopite, rutile, magnetite, ilmenite, apatite, monazite, and zircon have been observed within the sillimanite pseudomorphs and are in direct contact with each other. Corundum is frequently euhedral and has inclusions of fibrolite, indicating that its growth postdated the prograde transformation of andalusite to sillimanite. It is occasionally complemented by hematite of possible exsolution origin. Sapphirine is usually anhedral (skeletal) and intergrown with cordierite, sillimanite, spinel and phlogopite. Spinel from time to time forms symplectite with cordierite to partially replace sillimanite. The coronas surrounding the sillimanite pseudomorphs consist mostly of spinel and cordierite with minor corundum, sapphirine, phlogopite, magnetite, rutile, ilmenite, apatite, monazite, and zircon. The modes of occurrence of these minerals are basically the same as those within the sillimanite pseudomorphs apart from the lack of sillimanite in the coronas. The spinel−cordierite coronas are always enclosed by cordierite monomineral zones, which form the margin between the silica undersaturated and silica-saturated parts of the rock. The existence of sillimanite pseudomorphs after andalusite porphyroblasts positively point to the prograde P-T path followed by the rock from the andalusite to the sillimanite stability fields. The studied rock was divided into two different portions: silica-undersaturated sillimanite pseudomorphs and the silica-saturated matrix. Therefore, the main questions to answer are how and when the rock was divided into the two portions. The occurrence of phlogopite + quartz symplectites after orthopyroxene in the matrix suggests the following retrograde reaction:

$$\text{Orthopyroxene} + \text{K-feldspar} + H_2O \rightarrow \text{phlogopitz} + \text{quartz}. \tag{1.49}$$

However, reaction (1.50) is more plausible than reaction (1.49) for the following reasons.

$$\text{Orthopyroxene} + \text{granitic melt} \rightarrow \text{phlogopite} + \text{quartz} + \text{plagioclase}. \tag{1.50}$$

(1) Part of the plagioclase is found as grains containing fine-grained phlogopite and quartz, which are the result of reaction (1.50). (2) The modes of occurrence of quartz indicate that the quartz is of retrograde origin and, hence, there is a possibility that quartz was absent at the metamorphic peak. (3) K-feldspar is totally absent in the rock. It is improbable that all K-feldspar in the rock was used up by retrograde reaction (1.49). On the other hand, it is very credible that granitic melt produced by partial melting was removed from the rock, leaving no K-feldspar. Therefore,

it is suggested that granitic melt was present in the matrix at the thermal peak, which, in turn, suggests that the dehydration melting reaction (1.51) (reverse reaction of reaction 1.50) took place at prograde to peak metamorphic stages.

$$\text{Phlogopite} + \text{quartz} + \text{plagioclase} \rightarrow \text{orthopyroxene} + \text{granitic melt.} \tag{1.51}$$

The spinel—cordierite coronas surrounding sillimanite pseudomorphs are indicative of reactions between sillimanite and the matrix minerals, especially phlogopite.

$$\text{Sillimanite} + \text{phlogopite} \rightarrow \text{spinel} + \text{cordierite} + \text{K-feldspar} + H_2O. \tag{1.52}$$

However, reaction (1.53) is more likely than reaction (1.52) for the following reasons.

$$\text{Sillimanite} + \text{granitic melt} \rightarrow \text{cordierite} + \text{spinel} + \text{sapphirine} + \text{corundum.} \tag{1.53}$$

(1) K-feldspar, which should be formed by reaction (1.52), is completely lacking in the rock. (2) Dehydration melting to produce granitic melt is thought to have occurred in the rock, as discussed above. (3) The presence of corundum in addition to spinel and cordierite in the coronas hint at a silica-undersaturated environment, which is possible to appear during partial melting of relatively silica-poor rocks. (4) The observed coronas are identical with those commonly observed around Al_2SiO_5 xenocrysts in andesitic to rhyolitic magmas. In short, mineral textures strongly point to the presence of granitic melt at the thermal peak in the rock. The absence of K-feldspar and the depletion in the granitic component indicate that the geochemical nature of the rock can be ascribed, at least partly, to partial melting and removal of the produced granitic melt. It confirms the prograde P-T path of the rock from the andalusite to the sillimanite stability fields. Furthermore, the mineral assemblages in Mg-Al-rich rocks put forward a subsequent isobaric cooling path. The observed mineral textures are most fittingly elucidated by the presence of granitic melt in the rock at the thermal peak and its removal (Hiroi et al., 2001).

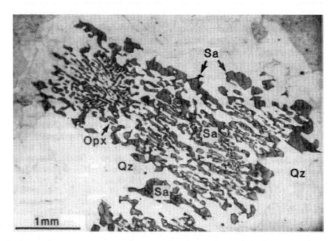

FIGURE 1.64   Sapphirine (Sa)-quartz (Qz)-orthopyroxene (Opx) symplectite. Modal proportion of orthopyroxene is generally less than 10% (Motoyoshi and Hensen, 1989 © www.schweizerbart.de).

Sapphirine—quartz—orthopyroxene symplectites in granulites of Mt. Riiser-Larsen, Enderby Land, Antarctica, have equant to rectangular shapes and a bulk composition equal to or approaching cordierite composition (Fig. 1.64). These symplectites characterize pseudomorphs after cordierite. The high Al_2O_3 concentration of the orthopyroxene (≤ 11.5 wt%) in the symplectites shows that they crystallized at near peak metamorphic conditions. Considering the shallow pressure versus temp. slope for the reaction cordierite $\rightarrow$ sapphirine + quartz + orthopyroxene, the symplectites formed in reaction to increasing pressure as the rocks move toward their thermal maximum. If taken in combination with the well-established isobaric cooling path for these granulites, an anticlockwise P-T path is implied. This explanation is at odds with crustal thickening (collisional) models for the development of these granulites and rather point to extensional tectonics, supplemented by massive heat input from mantle-derived magmas. Minor cordierite grains (0.5—3.0 mm) exhibiting multiple twinning are always surrounded by very fine-textured symplectitic intergrowths of cordierite + quartz. These symplectites formed late because they locally intrude the coarse-textured sapphirine—quartz—orthopyroxene symplectites. Where this happens, cordierite rims have crystallized between sapphirine and quartz. Volume analyses of the symplectitic intergrowths point toward a bulk composition indistinguishable from or come close to cordierite with $X_{Mg} = 0.82—0.83$. These observations lead to the inference that the symplectites are pseudomorphs after cordierite and have formed by its breakdown. The high Al_2O_3 content of the orthopyroxene suggests that the pseudomorphs formed under high-grade near-peak metamorphic conditions. The breakdown reaction of cordierite has a very flat dP/dT slope with cordierite on the low-pressure side. It has been proven that in the absence of CO_2, a reduction in water activity will likewise promote cordierite breakdown. Therefore, the sapphirine—quartz—orthopyroxene symplectites may have formed due to a pressure increase and/or a decrease in water activity. The textures suggest that peak metamorphic conditions were reached as pressure increased instead of pressure decreased. Alternatively, if cordierite had been stable before the peak metamorphic conditions as the interpretation of the symplectites here indicates, then an early high-pressure history is unlikely, irrespective of whether the symplectites formed by a pressure increase or by lowering of the water activity. Hence an anticlockwise P-T path for the Napier granulites is favored in combination with the well-established near isobaric cooling path. An extensional tectonic regime combined with massive heat input from the mantle, probably by underplating of mafic magmas, but devoid of substantial thickening of lower density crust, would best agree with the P-T history construed from the symplectites and the later postpeak development of the mineral assemblages of these granulites (Motoyoshi and Hensen, 1989).

Andalusite porphyroblasts are completely replaced as pseudomorphs by margarite-paragonite aggregates in aluminous pelites containing the peak mineral assemblage andalusite, chlorite, chloritoid, margarite, paragonite, quartz ± garnet, in an NW Iberia, Spain, contact area (Fig. 1.65). The absence of chloritoid resorption suggests either a pressure increase at constant reacting-system composition, or that its composition changed during retrogression at constant pressure, by becoming enriched in the progressively replaced andalusite porphyroblasts. Margarite—paragonite—muscovite pseudomorphs after chiastolitic andalusite are common in the aureole of the Boal granite. No andalusite remains were found in the pseudomorphs of the study by Martinez et al. (2004), but it can be found partially replaced in other parts of the aureole. The pseudomorphs are in the order of millimetres or, more often, centimeters in size and preserve the chiastolite shape and graphitic inclusion patterns. The matrix slaty cleavage can be seen within the finer-grained part of the pseudomorphs, which it also wraps around, forming pressure shadows. The pseudomorphs show that their coarse-grained part is formed by radial, fan-like margarite aggregates branching out from the pseudomorph rim. Some veinlets cut the aggregates and, in these, margarite laths grow at right angles to the walls. Paragonite is found in interstices between the margarite, and is uniformly dispersed in the finer-grained part of the pseudomorphs that are the inclusion-rich sections of the original chiastolite. The averaged pseudomorph mode in common rock and the andalusite-rich band is 1.85%, which in weight represent 1.83%. Area

(volume)—weight transformation was carried out assuming an average density of $2.8 \, g \, cm^{-3}$ for the mica, chlorite and quartz mixture in the matrix, and 3.58 for chloritoid, which only represents 2.85% in volume, hence the small volume—weight difference. It was presumed that diffusion during retrogression exceeded the pseudomorph-rich layer volume, as the andalusite in it became completely retrogressed, which means that the Al-rich layer did not act as a system insulated from the rest of the rock during retrogression. The assumption that it remained isolated would imply that the band was initially richer in the components required for its retrogression than the rest of the rock, or that depletion in the pseudomorph components was larger in the band, which is neither credible nor observed in the SEM images. Two assumptions were made for calculating porphyroblast retrogression; one was that a fluid phase was present during retrogression to account for the matter transport between the current phases responsible for porphyroblast replacement, the other was that Al was immobile in the porphyroblast and in the matrix. This model for retrogression should explain several observations such as (1) no andalusite left in the pseudomorph, i.e., the replacement was complete all over the rock, there being no differences in andalusite replacement in the andalusite rich band and in the rest of the rock; (2) chloritoid was not resorbed in the retrogression, and even limpid idiomorphic rims after turbid xenomorphic cores were found in matrix chloritoid; (3) the chloritoid mode appears comparable over the rock, with the anticipated exception of its increase in the quartz-absent sections of the tectonic banding; (4) pseudomorphs are rich in margarite and paragonite and nearly depleted in muscovite; (5) chloritoid composition must be Fe rich, and it appears to be the same throughout the rock; (6) chlorite remained stable, together with chloritoid. The model suggests that a real temperature-composition path should be a strongly curved one, suggesting a small temperature change as the main change of importance in

FIGURE 1.65 Hand sample showing chiastolite pseudomorphs totally replaced by muscovite—paragonite—margarite. The andalusite-rich layer has been considered to contain c. 27% of andalusite (Martinez et al., 2004).

composition from andalusite into pseudomorph had taken place at the andalusite-out isopleth, near 450°C. Throughout the retrogression, as the path pass through the four-variant field chloritoid—chlorite—muscovite-margarite—paragonite, there is a progressive increase of margarite and paragonite, which are incorporated into the pseudomorphs, and a decrease in quartz. At the same time, the matrix is being depleted in these phase components. As the reacting system that fulfills the chloritoid mode constraint should comprise the changing pseudomorph, the residual composition should be characterized by the depleted matrix. The former andalusite porphyroblasts became a preferential nucleation site for margarite, favored by the same Al/Si ratio in andalusite and margarite, and, hence, produce a mainly Ca-rich composition. The pseudomorph formation infers matter transport across the matrix toward the former andalusite porphyroblasts, and, assuming Al-immobility, this transport should involve Ca, Na, K, and Si, presumably by way of an intergranular H_2O-rich fluid. The preferential nucleation of margarite and paragonite in the pseudomorph, and its scarcity in muscovite, sug-

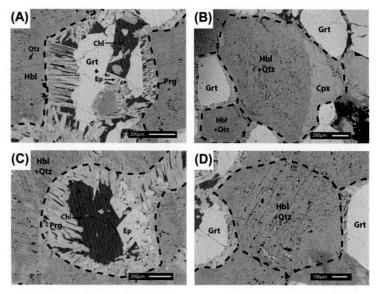

FIGURE 1.66   (A and B) Scanning electron microscopy (SEM) back-scattered images with location of minerals and representation of garnet and clinopyroxene shapes with *dashed lines* located close to the hydration front. (C and D) Back-scattered images with location of minerals and representation of garnet and clinopyroxene shapes (*dashed lines*) for a complete replacement (Centrella et al., 2015).

gests some kind of coupled Ca and Na transport, because margarite and paragonite increase amounts upon retrogression and muscovite diminishes and this supply seems directed to the prospective pseudomorph site. The balanced mole transfer for andalusite retrogression into the pseudomorph, which fulfills the pseudomorph composition, not including FeO and MgO, is

$$25.95 \text{ andalusite} + 9.72 \text{ } H_2O + 15.21 \text{ } SiO_2 + 6.32 \text{ } CaO + 3.69 \text{ } Na_2O + 0.75 \text{ } K_2O = 6.32 \text{ margarite} \\ + 7.38 \text{ paragonite} + 1.51 \text{ muscovite}$$
$$(1.54)$$

At constant Al, the 27.78 current Al_2O_3 moles in the pseudomorph, are the same as those in the original andalusite, and hence are used as a reactant, after subtracting the small amount (27.7 − 1.8 = 25.95) that is not linked to mica but to the small amount of chloritoid and chlorite present in the pseudomorph. CaO, Na_2O, and K_2O coefficients are the moles in the current pseudomorph that must have been gained during the replacement to produce the margarite, paragonite and muscovite. H_2O and SiO_2 coefficients are the amounts of these needed to produce the mica; they should be the same as those in the pseudomorph, once having subtracted the amounts in chloritoid and chlorite, and the amount linked to the original andalusite ($SiO_2 = Al_2O_3$) (Martinez et al., 2004).

The Precambrian granulite facies rocks of Lindas Nappe, Bergen Arcs, Caledonides of W. Norway were partially hydrated at amphibolite and eclogite facies conditions. However the replacements of garnet and of clinopyroxene are pseudomorphic so that the grain shapes of the garnet and clinopyroxene were preserved after they were fully replaced (Fig. 1.66). The textural evolution during the replacement of garnet by pargasite, epidote, and chlorite and of pyroxene by hornblende and follows what is expected for a coupled dissolution—precipitation mechanism. The element losses and gains in replacing the garnet are approximately balanced by the opposite gains and losses associated with the replacement of clinopyroxene. The coupling between dissolution and precipitation on both the grain and whole rock spatial scale preserved the volume of the rock during the hydration process. The sharp reaction interfaces between parent and product phases and the lack of any chemical zoning within the parent phases confirm that the replacement mechanism in both cases is by interface-coupled dissolution—precipitation (Putnis, 2009b; Putnis and Austrheim, 2013). What is notable however is that the preservation of the texture requires the transfer of elements from the dissolving clinopyroxene to the reaction that pseudomorphs the garnet and vice versa, i.e., the garnet needs to dissolve to provide the reactants for the clinopyroxene to hornblende + quartz reaction, instead of element diffusion within a single crystal through a reaction front (Fig. 1.67). However, preservation of texture during fluid—rock interaction is not unusual, although rarely remarked upon (Centrella et al., 2015).

The replacement of anhydrous by hydrous minerals is generally used to assume infiltration of a rock by H_2O. The Slate Creek complex is an early Jurassic arc ophiolite in the central portion of the western metamorphic belt, northern

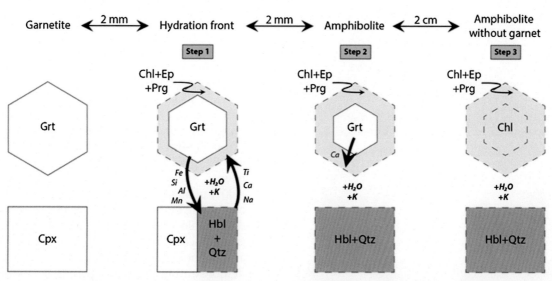

FIGURE 1.67　Schematic diagram illustrating element exchanges for the garnet and clinopyroxene replacement as a function of the location across the garnetite–amphibolite transition (Centrella et al., 2015).

Sierra Nevada, California. The ophiolitic pseudostratigraphy is preserved in a west-to-east sequence of metamorphosed ultramafic, plutonic, and volcanic protoliths (Edelman et al., 1989). Mineral assemblages and amphibole compositions indicate that metamorphic grade increases from subgreenschist-greenschist facies in the northeast to upper greenschist facies in the southwest (Fagan and Day, 1994). The textural transition from relict clinopyroxene to uralite is well preserved in the canyon of the South Fork of the Feather River. The transition can be observed in slates containing the assemblage: quartz, albite, chlorite, epidote, titanite, ±white mica, ±amphibole, ±rare pumpellyite. The clinopyroxene exhibits four textural stages of alteration: (1) clinopyroxene, (2) clinopyroxene + chlorite ± amphibole, (3) amphibole + chlorite, and (4) amphibole. The reaction path inferred from textural observations shows an early hydration step followed by a later dehydration step. Textural stage 1 corresponds with the original situation, where no clinopyroxene has been consumed. At textural stage 2, clinopyroxene was replaced primarily by chlorite with minor amphibole. A simple model based on these observations indicates that consumption of chlorite in the rock matrix released more water than required to produce amphibole pseudomorphs after clinopyroxene (also known as uralite). Thus the net flux of H_2O during uralitization of the greenschist facies metavolcanic rocks was outward. These results indicate that uralitization in these metavolcanic rocks resulted from heating rather than whole-rock hydration, and that natural mass fluxes may have been counter to fluxes inferred from textural evidence alone (Fagan and Day, 1997). This work indicates that the formation of pseudomorphic amphibole after clinopyroxene (uralitization) in low-grade metavolcanic rocks can be a net dehydration process.

The pseudomorphic replacement of chlorapatite by hydroxy-FAP is associated with metasomatism of the Ødegården metagabbro (Bamble, south Norway). Primary fluor-chlorapatite in the protolith gabbro is nonporous and homogenous in composition with an F-content up to 3.5 wt% and Cl-content up to 2 wt%. Metasomatism changed the apatite during multistage replacement reactions: (1) magmatic apatite was transformed to chlorapatite with a Cl-content up to 6.8 wt% during scapolitization and (2) a secondary pseudomorphic replacement reaction that transformed chlorapatite to porous hydroxy-FAP with only minor Cl is associated with albitization. The interface is sharp on a nanoscale, and the crystallographic orientation of the apatite is preserved. These observations, together with the porosity formed in the hydroxy-fluor-apatite, points to a replacement mechanism by interface-coupled dissolution–reprecipitation. The fluor-chlorapatite has a homogenous composition and is nonporous on the micron scale. In the scapolitized metagabbro, apatite is more prevalent and individual crystals display complex compositional zoning, identified as regions of chlorapatite and hydroxy-FAP. On textural grounds this has been interpreted as the partial replacement of chlorapatite by hydroxy-FAP. The chlorapatite often occurs in the central parts of the crystals, but also as irregular patches, sometimes along the rim, or into the crystal cores. The interface between chlor- and hydroxy-FAP is again sharp on a micron scale (Fig. 1.68). The apatite + phlogopite veins contain coarse apatite crystals. The apatite here exhibits similar textural and compositional relationships in the veins as in the scapolite metagabbro: The apatite is constituted by chlorapatite in the core of the crystals, whereas hydroxy-FAP occurs along the rims. In addition, chlorapatite shows a transformation to hydroxy-FAP related to microcracks into the crystals. The replacement occurs noncontinuously along the microcracks, and tends to focus around the fracture tips and in the dilational bridges occurring among overlapping segments of "en-echelon" fractures. As in the scapolite metagabbro, hydroxy-FAP has a high microporosity and pores

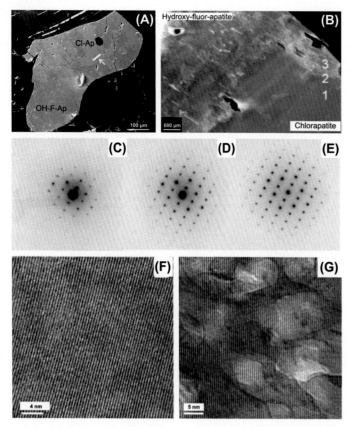

FIGURE 1.68 Transmission electron microscopy (TEM) investigation of a selected apatite grain. (A) FIB-scanning electron microscopy (SEM) image with a clear contrast between the chlor- and hydroxyfluor apatite phases (B) SEM dark field image. (C–E) Diffraction pattern across the interface in (B). Positions include (1) chlorapatite phase, (2) the interface, and (3) the hydroxyfluor apatite phase. (F and G) High-resolution images of the chlorapatite and hydroxyfluor apatite phase, respectively (Engvik et al., 2009).

occasionally occur in trails at the interface. Tiny inclusions of monazite and xenotime can be found in the hydroxy-FAP. The chlorapatite shows smooth diffraction contrast throughout, whereas the hydroxy-FAP has large pores as well as complex contrast which are most likely due to the presence of nanopores, as shown in the high resolution TEM image (Engvik et al., 2009).

Octahedral graphite aggregates are present in garnet pyroxenites from the Ronda peridotite massif, southern Spain. In the aggregates, basal {0001} planes of graphite are parallel to {111} of the octahedra, demonstrating that the graphite is pseudomorphic after a cubic mineral (Fig. 1.69). The aggregates contain clinopyroxene inclusions that have a morphology atypical of a monoclinic phase, demonstrating that the precursor diamond has imposed a cubic symmetry. The largest graphite octahedron so far found at Ronda is equivalent to an eight carat diamond. The high abundance of graphite implies original diamond content over 15%. This concentration is $\approx 10^7$ carats per ton, which is $\approx 10^5$ times richer than kimberlite and lamproite deposits and comparable to the richest diamondiferous eclogite xenoliths (Gurney, 1989). Graphitized diamonds at Ronda and at Beni Bousera (northern Morocco) establish that mantle-derived material is brought to the surface from great depths by a method other than kimberlitic volcanism. The graphite garnet pyroxenites at Ronda have undergone substantial recrystallization. The relict minerals (clinopyroxene + garnet + quartz) and bulk rock chemistry are consistent with formation from an eclogite oversaturated in silica. The following phase changes are thought to have taken place at mantle depths: diamond → graphite and coesite → quartz. Slab detachment resulted in relatively hot and buoyant asthenosphere replacing the cold subducting slab and caused significant uplift and heating of the overlying continental crust. It has been envisaged that peridotite emplacement was a multistage process with initial uplift from the diamond stability zone and diamond graphitization associated with continental collision at ≈ 85 Ma. However, the major tectonic displacement (≈ 100 km) was associated with slab detachment at −2 0 Ma. Initial uplift caused by the induced flow of the asthenosphere was followed by nappe emplacement as the uplifted continental lithosphere underwent extensional collapse. Rapid cooling and subsidence of the thinned lithosphere over the past 20 m.y. has produced the Alborán Basin (Davies et al., 1993). A detailed mineralogical and textural study of two pyroxenites from the Beni Bousera massif, a garnet clinopyroxenite (GP) and a garnet clinopyroxenite containing graphite pseudomorphs after diamond (GGP), indicates a strong metamorphic overprint associated with massif exhumation. GGP is characterized by graphite aggregates (2–5 mm) with an octahedral morphology which has been taken as evidence for pseudomorph after diamond (El Atrassi et al., 2011; Pearson et al., 1989, 1995, 1991; Slodkevich, 1980). These graphite aggregates conspicuously host large silicate inclusions (250–1000 µm) in their core. These inclusions have a comparable microstructure as the pyroxene porphyroclasts. They consist of clinopyroxene and orthopyroxene lamellae and, possibly, of secondary garnet (Fig. 1.70). Pearson et al. (1989) first observed the existence of clinopyroxene and garnet in the graphite pseudomorphs. These inclusions, the mineralogy of which is consistent with the primary bimineralic nature of the host GGP (clinopyroxene + garnet) were construed as trapped in the former diamond (Pearson et al., 1989). Indeed, these authors used, in addition to morphological considerations, the chemical similarity (major and trace elements)

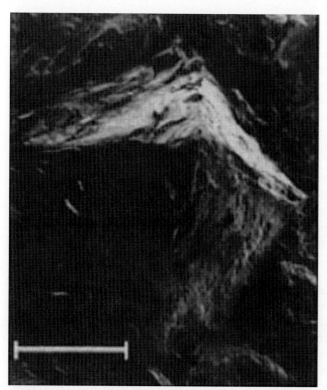

FIGURE 1.69 Scanning electron microscope (SEM) image of upper half of graphite octahedron in garnet pyroxenite from Ronda peridotite massif. Scale bar represents 1.36 mm (Davies et al., 1993).

between these inclusions and their equivalents in the adjacent host-rock as proof for a cogenetic relationship between GGP and diamonds (El Atrassi et al., 2013).

Globular, platy, and finely dispersed phyllosilicates of chloritic composition can be found in the outer contact of dike with glauconite-bearing rocks from the lower Ust'-Il'ya Formation at the Kotuikan River 2.5 km upstream the Il'ya River which is part of the Anabar Uplift, Siberia. It has been shown that the globular Al-glauconite was replaced by pseudomorphs of mixed-layer Mg- and Fe-bearing chlorite—berthierine (containing 5% berthierine layers in this zone). In order to understand the mechanism of chloritization of glauconite globules, it should be noted that phase transformations nearly did not affect the size, shape, and character of the globule surface. This shows that, first, chloritization was not accompanied by significant changes in globule volume; i.e., the amount of building material accumulated by the globules approximately equaled to the amount removed from globules. Second, the mentioned features indicate that phase transformations were not accompanied by intense dissolution and growth of crystals from fluids. Otherwise, the globules would have uneven (corroded) surface and their volume would change. Therefore it has been suggested that the mineral formation environment was not extremely aggressive and chloritization was slow enough to allow the migration of cations and anions inside and outside the globules without significant morphological changes. Another important feature of the considered phase transition may be related to the fact that micas and chlorites mainly consist of 2:1 layers. It was originally thought that mica chloritization was a solid-state reaction with the preservation of major structural elements of the micaceous matrix (Drits et al., 2001). This interpretation is now been thought as being erroneous and more likely an interface-coupled dissolution—reprecipitation process along a progressing through a thin fluid film has been responsible for the reaction.

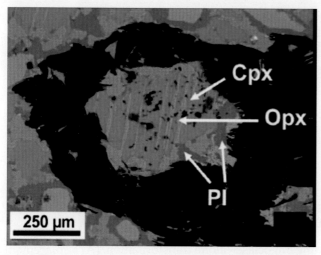

FIGURE 1.70 Details of a silicate inclusion in graphite aggregates from GGP from the combination of X-ray elemental maps (Cpx = clinopyroxene, Opx = orthopyroxene, Pl = plagioclase) (El Atrassi et al., 2013).

Large-scale metasomatic albitization in the albite terranes of the Bamble sector of southeastern Norway has affected both mafic and granitic lithologies. In partially metasomatized tonalite, the albitization fronts advanced normal to fractures and can be recognized in the field by a distinctive reddening of the rock where original plagioclase crystals have been replaced by albite. Although albitization occurs on a regional scale, the mechanism responsible takes place on the scale of separate mineral grains. In a more general sense, the replacement of one feldspar by another is ordinarily related to fluid—rock interaction and metasomatism. A distinctive feature of replaced feldspars is the development of turbidity due to the formation of porosity. The initial albitization occurs along thin veins filled by albite and calcite. The fractures control the replacement of the tonalite to the typical mineral assemblage of albitite: albite + calcite + chlorite + epidote + rutile. The transformation of oligoclase to albite happens over a few centimeters in the contact region between the light gray tonalite and the red regions. In SEM BSE images, the parent plagioclase (oligoclase $An_{22}Ab_{77}Or_1$) seems to be homogeneous and free of any apparent porosity. The transformation to albite ($An_3Ab_{95}Or_2$) occurs at sharp interfaces on a microscale and along tiny fractures cutting through the oligoclase part of the crystals, which can be found as trails of pores or microveins filled by albite (Fig. 1.71). The albite is prominently porous and occurs with abundant crystals of very fine-grained white mica (up to 10 mm length) and fine inclusions of Fe oxide (about $1-2$ mm), together with smaller amounts of calcite, epidote and chlorite. Locally, very fine grains of chlorite and calcite are present. The sharp oligoclase—albite replacement interface, visible in the BSE image, is accompanied by a sharp chemical transition as the feldspar composition shows an abrupt change from $An_{22}Ab_{77}$ to An_3Ab_{95}. Close to the albite-filled vein, the rock is completely transformed to albitite, where the feldspar occurs as pure reddened albite (An_1Ab_{99}) with a high microporosity. The partially replaced plagioclase exemplifies the most important characteristics of mineral—replacement reactions induced by fluids: (1) the replacement interface is sharp on a nanoscale, with little proof of any significant solid-state diffusion between parent and product phases, and the crystallographic orientations of oligoclase and albite are parallel within no more than one degree. The sharpness of the chemical reaction front is unmistakable. (2) The reaction product is porous. The "porosity" is the volume occupied by the fluid phase and is created at the reaction interface anywhere the volume of the parent mineral dissolved is less than the volume reprecipitated. This volume deficit is reliant on not only on the differences in molar volume between the solid phases, but also on relative solubilities of the solids in the fluid at the interface. Thus it is conceivable to generate porosity even where the molar volume of the product solid is higher than that of the parent solid. (3) The original external volume occupied by the original plagioclase crystals in the rock as well as their crystallographic orientation are preserved during the albitization process. (4) other phases may coprecipitate during the albitization reaction. In this example, precipitation of very fine particles of Fe oxide is associated with the albitization, causing the red coloration in the rock. These are all features common to the interface-coupled dissolution—reprecipitation mechanism that has been proposed to be a general mechanism of reequilibration of solids in the presence of a fluid phase (Putnis, 2009b; Putnis and Austrheim, 2010, 2013; Putnis and Putnis, 2007). In this mechanism, at the reaction interface, the orthoclase begins to dissolve, resulting in an interfacial fluid that is supersaturated with respect to the product phase (albite), which reprecipitates. The crystallographic

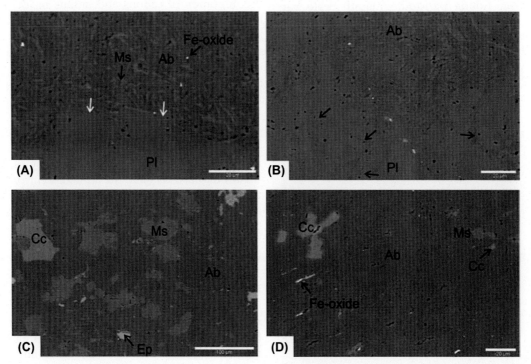

FIGURE 1.71   Back-scattered electron (BSE) images illustrating oligoclase—albite transformation. (A) Replacement of oligoclase (An22Ab77) by albite (An2Ab95). The replacement interface (*white arrows*) is sharp. Note that oligoclase appears texturally homogeneous without pores and that the albite part of the crystal has a high microporosity. Light, very fine-grained laths in the albite are white mica, whereas *small bright spots* represent particles of Fe oxide. Partly transformed tonalite. Scale bar is 20 μm. (B) Replacement of oligoclase by albite. The replacement interface is sharp, but fractures visible as trails of micropores and albite-filled microveins (*arrows*) cut through the oligoclase. Partly transformed tonalite. Scale bar is 20 μm. (C) Albite containing coarser inclusions of calcite, white mica and epidote. Albitite. Scale bar is 100 μm. (D) Porous albite, including calcite and white mica inclusions, and small *bright spots* of Fe oxide. Albitite. Scale bar is 20 μm (Engvik et al., 2008).

orientation of the orthoclase is preserved at first owing to epitaxial nucleation of the albite on the dissolving surface of the orthoclase; subsequently, albite can either grow upon oriented nuclei or form new oriented nuclei. The mechanism calls for the rates of dissolution and reprecipitation to be coupled at the interface. The creation of porosity is a key feature of the mechanism and allows fluid infiltration to the reaction interface, through the product phase. It is obvious that fractures act as fluid channels on the outcrop scale during the albitization, supplying fluid and matter for the reddening of the rock and to allow the new mineral assemblage to form. The microtextures described from the abilities, where the replacement minerals are found along microveins, accentuate the significance of fluid circulation along microcracks as transport channels to supply elements for mineral—replacement reactions. However, microcracks alone cannot account for the interaction of the fluid with every part of the plagioclase crystal that is ultimately completely replaced. Porosity, and hence permeability, generated by the interface-coupled dissolution—reprecipitation mechanism, provides a more pervasive mechanism of fluid infiltration. The BSE images show that both microcracks and porosity generation guarantee contact between the reactive fluid and the parent mineral. The presence of inclusions of white mica closely associated with the albitization indicates that Al^{3+} is released from the orthoclase. The evidence supporting a cogenetic origin for the albite and its inclusions is primarily textural: the very fine distribution of white mica in the albitized parts of the plagioclase and the fact that, in the TEM images, white mica is closely associated with porosity in the albite. The formation of a white mica always requires the introduction of K into plagioclase (Que and Allen, 1996), an observation that is frequently made when describing alteration in rocks, but whose origin is rarely discussed. The breakdown of biotite to chlorite in tonalite is one potential source of K, although as albitization requires the introduction of Na by fluids external to the tonalite, it is quite conceivable that such fluids could also have contained K (Engvik et al., 2008).

Aragonite, a high-pressure polymorph of calcite, is the stable calcium carbonate mineral during metamorphism at blueschist to eclogite facies conditions. Nevertheless, aragonite is rare in high-pressure marbles because it transforms to calcite during exhumation. Although marbles on the island of Syros (Cyclades, Greece) contain high-pressure mineral assemblages consistent with the blueschist to eclogite facies metamorphism recorded by associated mafic rocks, no aragonite has been identified there. Because these rocks show evidence of an incomplete greenschist facies overprint, the absence of aragonite is not surprising. However, numerous rod-shaped calcite crystals have been observed in the blueschist to eclogite facies marbles of Syros, Greece. These rods show a shape-preferred orientation, and the long axes of the rods are oriented at a large angle to foliation. The crystals also have a crystallographic-preferred orientation: calcite c-axes are oriented parallel to the long axes of the rods. Based on their chemical composition, shape, and occurrence in high-pressure marbles, these calcite crystals have been interpreted as topotactic pseudomorphs after aragonite, which developed a crystallographic-preferred orientation during peak metamorphism. This interpretation is consistent with deformation of aragonite by dislocation creep, which has been observed in laboratory experiments (Rybacki et al., 2003) but has not been previously reported on the basis of field evidence. Subsequent to the high-pressure deformation of the aragonite marbles, the aragonite recrystallized into coarse rod-shaped crystals, maintaining the crystallographic orientation developed during the deformation stage. During later exhumation, aragonite reverted to calcite. Experiments on the transformation of aragonite to calcite due to a decrease in pressure offer constraints on the significance of the pseudomorph textures observed in the Syros marbles. Although the results of experimental studies on this transformation during deformation (Gillet et al., 1987; Snow and Yund, 1987) are difficult to reconcile, experimental studies on the transformation under static conditions have highlighted the significance of topotaxy (which preserves crystal lattice orientations) during the phase transformation. Both Brown et al. (1962) and Carlson and Rosenfeld (1981) observed that a major fraction of the new calcite crystals had c-axes parallel to the original aragonite c-axes (parallel to the long axis of acicular crystals). Other experiments indicated that calcite that has replaced aragonite by topotaxy is generally biaxial with a low 2V (Boettcher and Wyllie, 1967). The biaxial character of most calcite grains strengthens the interpretation that the acicular calcite textures here represent pseudomorphs after aragonite (Brady et al., 2004).

Large (3 cm) euhedral trapezohedra composed of single crystals of analcime or aggregates of pumpellyite occur in stratigraphically adjacent phonolitic lavas from Aghda, Central Iran. The trapezohedra contain concentrically arranged inclusions of plagioclase, pyroxene, titanomagnetite, and apatite that are fresh in the analcime but extensively altered in the pumpellyite (Fig. 1.72). The analcime crystals are remarkably homogeneous in composition and are interpreted as having formed by ion-exchange pseudomorphous replacement of primary leucite. In the lower phonolite, however, the analcime became unstable and was replaced by pumpellyite, which has a wide compositional range, reflecting variable input of Fe, Mg, Ca, and Al from the precursor analcime and alteration of inclusions. The occurrence of analcime and pumpellyite at Aghda is indicative of zeolite facies conditions with low f_{CO_2} and H_2O activity and probably high P_{fluid}. Analcime trapezohedra appear optically continuous, suggesting that they are single crystals. The pumpellyite trapezohedra, however, are composed mainly of an aggregate of randomly arranged colorless pumpellyite flakes up to 0.25 mm long, but most grains are <0.05 mm in size. Prehnite is rather rare in the trapezohedra and the main chemical variation relates to minor Fe—Al substitution. The compositions are typical of prehnite produced during hydrothermal alteration and low-grade metamorphic facies. All plagioclase inclusions in the pumpellyite trapezohedra have been altered to albite, sericite and kaolinite. The

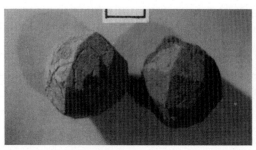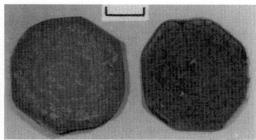

FIGURE 1.72   Left euhedral trapezohedra of pumpellyite (left) and analcime (right). Bar scale is 1 cm long. Right photograph of polished sections through centers of pumpellyite (left) and analcime (right) showing inclusions arranged in a well-developed concentric pattern. Bar scale is 1 cm long (Carr et al., 1996).

ease, rapidity and selectivity of the replacement process are evidenced by the lack of effect on the other phases included in analcime. Formation of analcime as an intermediate phase rather than direct replacement of primary leucite by pumpellyite is envisaged due to the apparent ease and rapidity of analcime replacement of leucite. The preferential replacement of analcime by pumpellyite in a specific phonolite lava may reflect differences in permeability. The most important variables in determining the composition and identity of the pseudomorphous replacement mineral appear to have been permeability, fluid chemistry and P_{fluid} (Carr et al., 1996). Cassedanne and Menezes (1989) reported on a new occurrence of pseudoleucite pseudomorphs from the Rio das Ostras area (Brazil), where this mineral was found in a highly weathered, lenticular dike. The alteration has left the original crystal shape unchanged. The trapezohedron is the only form observed. Single crystals represented ~90% of all the pseudomorphs found, the other being twins and groups. The alteration of the original leucite was probably due to a late hydrothermal stage during dike emplacement. It produced mainly Al_2O_3 (including gibbsite) and kaolinite containing impurities leached from other minerals. The pseudoleucites are composed of aggregates of orthoclase in irregular grains and, more often, radiating prisms. Interstitially there is a comparatively small amount of nepheline, usually as irregular patches but rarely as prismatic crystals. The cores of the original leucite crystals are filled with analcime or, occasionally, a weakly doubly refracting zeolite as yet unidentified. In some cases the pseudoleucites are zoned, containing a potassium-rich alkali feldspar and exhibiting a trapezohedral form. In outcrops where kaolinite is the main constituent, the pseudomorphs are whitish to pale yellow in color with somewhat rounded edges. Such samples are easily destroyed or damaged by immersion or soft washing. On the other hand, specimens in which aluminum hydroxides are predominant (pinkish to pale beige in color) are stronger and will withstand washing. Some pseudomorphs have pitted faces or show a thin lustrous coating of an unidentified mineral.

Pseudomorphs after lawsonite from the Glockner nappe (south-central Tauern Window, Austria), a metamorphic sequence of metasediments intercalated with metavolcanics of mid-ocean ridge basalt origin, were described by Gleissner et al. (2007). The basalts were metamorphosed in blueschist facies, as shown by the pseudomorphs after lawsonite and relics of glaucophane, and then overprinted at greenschist-facies conditions. The foliated fabric suggests late syndeformative greenschist facies formation of the pseudomorphs. The pseudomorphs show the mineral assemblage Albite + Epidote + Phengite + Chlorite + Calcite + Titanite + Actinolite ± Quartz ± Apatite. The shape of the pseudomorphs varies from rhombic or rectangular-euhedral to rounded and not very well preserved. The dark areas in the pseudomorphs are epidote rich. No preferred orientation of the pseudomorphs was observed (Fig. 1.73). The pseudomorphs contain albite, epidote, white mica and smaller amounts of chlorite, calcite and titanite. Accessory phases are amphibole, quartz, pyrite, and apatite. The rock foliation continues into the pseudomorphs. The internal foliation is outlined by discontinuous thin layers of epidote and shows a straight lined to sigmoidal pattern. The confirmation for former lawsonite (the pseudomorphs) in these metabasalts was ascribed to a prograde HP-LT blueschist facies metamorphism. The small glaucophane relics in the pseudomorphs from the Rauh Kopf confirm this elucidation. From the high pressure-low temperature (HP-LT) assemblage lawsonite + glaucophane and the present mineral assemblage, indicating greenschist facies conditions for the last metamorphic over print, a P-T evolution was inferred that exceeded the thermal stability of lawsonite between 380°C/7 kbar and 440°C/12 kbar. Epidote marks the pseudomorph foliation and is the phase that was able to incorporate Ca and Al of the lawsonite breakdown reaction (note that feldspar is present as pure albite). The internal foliation therefore cannot be inherited from a lawsonite porphyroblast that overgrew an earlier foliation. These details and the fact that the internal fabric is the continuation of the matrix foliation (which is composed of greenschist facies minerals) led to the conclusion that the foliation within the matrix and the pseudomorphs were formed at the same time. The apparent paradox between the internal foliation and the euhedral lawsonite shape suggests the syndeformative decomposition of lawsonite although the deformation was not sufficient to destroy the lawsonite shape. This was

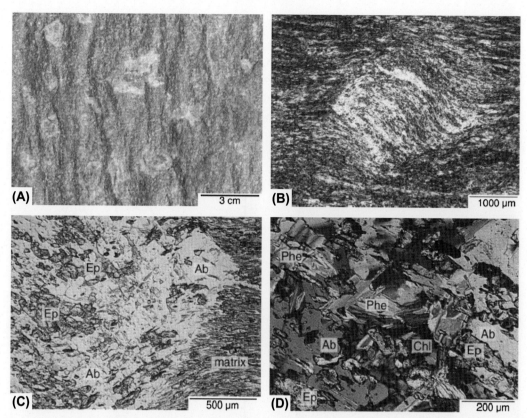

FIGURE 1.73 Photographs of greenschist from the Glockner nappe. (A) Hand specimen shows lawsonite pseudomorphs. (B and C) Thin section photomicrographs of foliated greenschist with pseudomorph after lawsonite (plane-polarized light). (D) Pseudomorph phases Ab = albite, Ep = epidote, Phe = phengite, Chl = chlorite (crossed polars) (Gleissner et al., 2007 © www.schweizerbart.de).

interpreted as decomposition of lawsonite at the end of major penetrative deformation and fabric formation. The sigmoidal foliation pattern found in some pseudomorphs suggests a slight rotation of lawsonite during decomposition instead of destruction by shearing. The difference in metamorphic assemblage and the less foliated fabric suggest an actinolite consuming and chlorite and calcite producing reaction. This mechanism requires the input of CO_2 rich fluids, which likely came from the surrounding calcareous Bündnerschists.

Pseudomorphs in amphibolitized eclogites of the Nevado-Filabride Complex, Betic Cordilleras, Spain, were described by Gomez-Pugnaire et al. (1985). Three different textural and mineralogical types can be recognized. All contain a combination of the minerals kyanite (ky), margarite (ma), paragonite (pa), epidote, and quartz (qz). Aggregates of calcite + epidote + paragonite + albite have been interpreted as pseudomorphs of lawsonite (law) porphyroblasts by Puga (1977). Aggregates of paragonite + margarite + epidote + kyanite, in some cases with coronitic textures, were deduced to be break-down products of lawsonite. Type A consists of large crystals of kyanite (length 1—2 mm), corroded by intersected white mica (Fig. 1.74A). Occasionally only a few relics of kyanite are preserved. Large kyanites are optically continuous, small crystals may be either optically continuous or discontinuous. Infrequently small amounts of carbonate occur in the center of the pseudomorphs: in one case additional chlorite was observed. Similar textures of the transformation of kyanite were observed in the surrounding metapelites of the Sierra de Baza. This suggests ionic reactions like:

$$2ky + Ca^{2+} + 2H_2O \rightarrow 1ma + 2H^+ \tag{1.55}$$

$$3ky + 2Na^+ + 3SiO_2 + 3H_2O \rightarrow 2pa + 2H^+ \tag{1.56}$$

where ky = kyanite, ma = margarite and pa = paragonite, The reactions are balanced to constant Al with a mica formula $XY_2Z_4O_{10}(OH)_2$. Although these mineral transformations are written as ionic exchange reactions, they do not require a large-scale metasomatism. During breakdown of omphacitic pyroxene into amphibole and albite near to the kyanite crystals, Ca as well as Na and Si would be ionic species in a fluid phase. By this process the Ca^{2+}/H^+

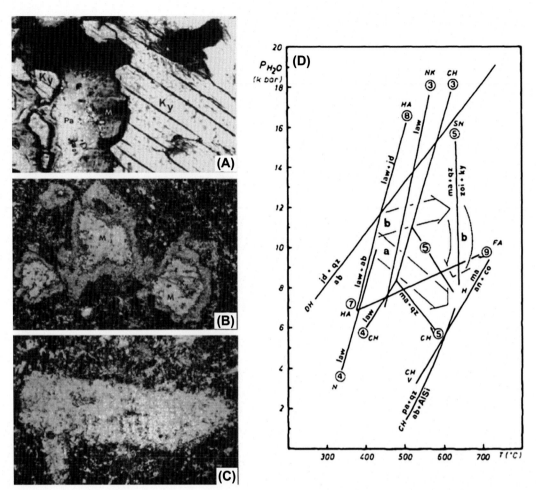

FIGURE 1.74 (A) Type-A pseudomorphs, detail: few relics of kyanite (Ky), surrounded by margarite (M) and paragonite (Pa). Margarite always occurs as rims around kyanite, and paragonite as rims around margarite (crossed nicols); real length of the picture is 1.2 mm. (B) Type-B pseudomorphs with white mica (M, margarite and paragonite), kyanite (Ky), and a coronitic rim of epidote (E), (crossed nicols); real length of the picture is 4.7 mm. (C) Type-C pseudomorphs: paragonite + quartz (low relief) and epidote (high relief), crossed nicols; real length of the picture is 4.2 mm. (D) Experimentally determined or calculated equilibrium curves and possible extreme P-T paths (A) and (B) for the development of the rocks. Encircled numbers refer to reaction numbers in the text [DH = Delaney and Helgeson (1978); CH = Chatterjee (1971, 1972, 1976); FA = Franz and Althaus (1977); HA = Heinrich and Althaus (1980); N = Nitsch (1973); NK = Newton and Kennedy (1963); SN = Storre and Nitsch (1974); H = Holland (1979); V = Velde (1971)]. For abbreviations and reaction numbers see the text (except co = corundum) (Gomez-Pugnaire et al., 1985 and references therein).

and Na^+/H^+ ratios of the fluid are considerably changed, which has a great impact upon the stability of the aluminum silicate. The textures of these pseudomorphs point to first margarite and then paragonite replacing kyanite, or paragonite replacing margarite. Type B exhibits a coronitic texture (width of the corona 2–2.5 mm). Their shape is characteristically rhombohedral and prismatic, mostly euhedral to subhedral. They have a core of white mica with small irregularly distributed and differently orientated crystals of kyanite, and a corona of epidote (Fig. 1.74B). In both cases A and B, paragonite and margarite are both present in the micaceous aggregates. Type-B pseudomorphs exhibit a more complicated mineralogy and texture. Their shape point to them being formed after lawsonite. However, no relics of lawsonite have been found. There are two different extreme possibilities to explain their formation from lawsonite (which are called as (1) and (2) in the following section): (1) In the first stage lawsonite was altered (more or less concurrently) according to the following reactions (1.57)–(1.59) at P-T conditions where these three reactions meet in an invariant point:

$$4law \rightarrow 2zoi + 1ky + 1qtz + 7H_2O \tag{1.57}$$

$$5law \rightarrow 2zoi + 1ma + 2qtz + 8H_2O \tag{1.58}$$

$$2zoi + 5ky + 3H_2O \rightarrow 4ma + 3qtz \tag{1.59}$$

where law = lawsonite, zoi = zoisite and qz = quartz. The pseudomorphs would then consist of margarite, quartz, kyanite and zoisite. It may seem doubtful that the P-T conditions were exactly at the invariant point, but it must be kept in mind, that this "invariant" equilibrium becomes at least bivariant due the Fe^{3+} content of epidote. In the later stage, margarite was altered into paragonite by the exchange reactions:

$$ma + Na^+ + Si^{4+} \rightarrow 1pa + Ca^{2+} + Al^{3+} \tag{1.60}$$

$$3ma + 4Na^+ + 2H^+ + 6SiO_2 \rightarrow 4pa + 3Ca^{2+} \tag{1.61}$$

where reaction (1.60) is balanced with respect to 1 "mica unit", and reaction (1.61) is balanced with respect to constant Al. These reactions can explain the absence of quartz in the pseudomorphs, which should be present based on reactions (1.57)–(1.59), and also the rim of epidote around the pseudomorphs. The process is comparable to the ionic reaction formulated above for pseudomorphs of type A. Ca^{2+} and Al^{3+}, released by this process, could form the monomineralic epidote corona of the pseudomorphs. Such monomineralic zones are typical of metasomatism and point to the high mobility of many components. (2) Lawsonite was converted into zoisite + kyanite + quartz only (not margarite) by crossing the boundary of reaction (1.57). In a later stage margarite was formed by a retrograde hydration reaction (1.59), and in a yet later stage [similar to possibility (1)], margarite was transformed into paragonite by high activity of Na^+. Type-C pseudomorphs can easily be explained by breakdown of the association lawsonite + albite, or lawsonite + jadeite, according to reactions (1.62) or (1.63):

$$4law + 1ab \rightarrow -1pa + 2zoi + 2qtz + 6H_2O \tag{1.62}$$

$$4law + 1jd \rightarrow 1pa + 2zoi + 1qtz + 6H_2O \tag{1.63}$$

where ab = albite and jd = jadeite. It is tacitly assumed that albite or jadeite were transported from the matrix to the lawsonite as components of the fluid phase in order to maintain the euhedral shape of the pseudomorphs. Type C quartz and epidote are generally associated with large flakes of intersecting paragonite (Fig. 1.74C). The replaced mineral was euhedral. They reach a length of 2–4 mm, and occur regularly in the metabasites. Very small relics of kyanite rimmed by margarite are found as inclusions in the large paragonite flakes.

The shape of the pseudomorphs C leaves open the possibility that the pseudomorphosed mineral was not lawsonite, but plagioclase. In this case an anorthite-rich plagioclase may have been transformed directly into the assemblage paragonite + zoisite + quartz:

$$4an + 1ab + 2H_2O \rightarrow 1pa + 2zoi + 2qtz \tag{1.64}$$

where an = anorthite. Note that this is a hydration reaction, whereas the above-mentioned reactions are dehydration reactions. This last reaction also necessitates that the assemblage paragonite + zoisite + quartz was the eclogite paragenesis, otherwise these pseudomorphs never passed through a stage of eclogite equilibrium without plagioclase. Starting from the assumption that the rocks were first metamorphosed at low temperatures/high pressures, this hypothesis is supported by the existence of relics of glaucophane and omphacite. The likelihood that the type-C pseudomorphs were formed straight from plagioclase can therefore safely be excluded, because this would necessitate a P-T path from low to intermediate temperatures/low pressures to P-T conditions near to the boundary of the paragonite + zoisite + quartz stability field. The pseudomorphs therefore are more likely to have formed after lawsonite, and the rocks must have passed through the lawsonite (and lawsonite + albite or lawsonite + jadeite) stability field, which is somewhere left of reactions (1.57), (1.58), (1.62), and (1.63) (Fig. 1.74D).

Polyphase mineral aggregates (PMAs) containing clinozoisite + kyanite + quartz ± chlorite ± paragonite ± phengite have been observed inside garnet and in the matrix of talc-garnet-chloritoid schists from the Makbal ultrahigh–pressure complex in the northern Kyrgyz Tian-Shan (Fig. 1.75). These mineral textures are interpreted as pseudomorphs after lawsonite, the reconstruction of the compositions of PMAs of clinozoisite + kyanite + quartz shown to be consistent with lawsonite. Petrological study revealed that lawsonite was stable during the prograde stage to the UHP peak stage (P = 28–33 kbar and T = 530–580°C) and decomposed to the PMAs in the course of

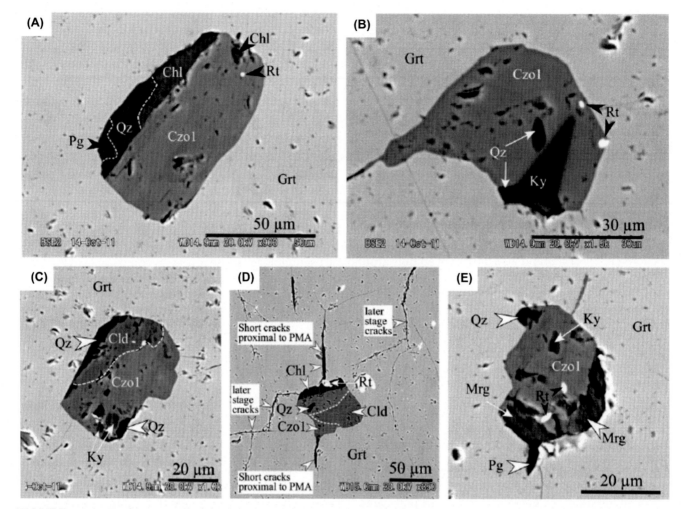

FIGURE 1.75 Back-scattered electron images showing the textures of the PMAs in garnet and in the matrix of the rocks: (A) prismatic PMA consisting of Czo1 + Chl + Qz + Pg; (B) PMA consisting of Czo1 + Ky + Qz; (C) PMA consisting of Czo1 + Ky + Qz + Cld; (D) prismatic shaped PMA consisting of Czo1 + Chl + Qz + Cld; short cracks filled by chlorite are present around PMA and these cracks are not connected to later-stage cracks; (E) margarite replacing a PMA consisting of Czo1 + Ky + Qz (Orozbaev et al., 2015).

isothermal decompression around P = 16–20 kbar and T = 510–580°C. The textural features of prismatic shaped PMAs largely consisting of clinozoisite 1 + quartz + kyanite, clinozoisite 1 + chlorite + paragonite + phengite + quartz within garnet crystals, and clinozoisite 2 + chlorite + quartz in the matrix indicate the likelihood that these PMAs originated from earlier lawsonite. This is in line with accounts that such PMA assemblages with prismatic shapes are considered as breakdown products after lawsonite. Normally, lawsonite breakdown reactions do not occur in a closed system; most described pseudomorphs after lawsonite contain Na- and K-bearing phases (e.g., paragonite, albite, and phengite) and/or Fe- and Mg-bearing phase exemplified by chlorite. Therefore, it is usually challenging to recreate original lawsonite compositions from PMAs considered as pseudomorphs after lawsonite. In a study by Orozbaev et al. (2015), they reconstructed the original lawsonite compositions from the fine-grained PMAs (<100 μm) of clinozoisite 1 + kyanite + quartz preserved inside a garnet and thus circumvented interaction with other matrix phases. This texture is indicative for lawsonite breakdown according to the reaction:

$$4 \text{Lawsonite} = 2 \text{ Zoisite} + \text{Kyanite} + \text{Quartz} + 7\text{H}_2\text{O} \tag{1.65}$$

In contrast another sample did not contain kyanite. The main assemblages of the coarse-grained PMAs (100–600 μm) were clinozoisite 1 + chlorite + paragonite + phengite + quartz within the garnet and clinozoisite 2 + chlorite + quartz ± chloritoid in the matrix The principal minerals of the PMAs within the garnet point to lawsonite breakdown accompanied by interaction with other Na- (e.g., jadeite, omphacite, or glaucophane) and K-bearing (phengite) phases. Even though these Na-bearing minerals were not observed as inclusions in the garnet crystals, glaucophane was observed as a matrix phase in the first sample, which might point toward the following mineral reaction producing the PMAs with paragonite:

$$\text{Garnet} + \text{glaucophane} + \text{lawsonite} = \text{zoisite} + \text{paragonite} + \text{chlorite} + \text{quartz} + \text{H}_2\text{O} \qquad (1.66)$$

PMAs were found in the core of garnet crystals, suggesting the existence of lawsonite during the prograde stage. Therefore, the prograde path is qualitatively indicated by the inclusion phases of lawsonite (now pseudomorphosed), Mg-rich chloritoid, talc, chlorite, quartz, and rutile within the garnet crystals. The petrogenetic grid for the NCKFMASH ($\text{Na}_2\text{O-CaO-K}_2\text{O-FeO-MgO-Al}_2\text{O}_3\text{-SiO}_2\text{-H}_2\text{O}$) system of Wei and Powell (2006) shows formation of the garnet-chloritoid-talc assemblage at the cost of chlorite, lawsonite, and quartz around P = 20−25 kbar and T = 500−560°C, according to the following reaction:

$$\text{Chlorite} + \text{lawsonite} + \text{quartz} = \text{garnet} + \text{chloritoid} + \text{talc} + \text{H}_2\text{O} \qquad (1.67)$$

This reaction provides a pressure limitation for the prograde stage, however the temperature must have been below peak conditions based on the prograde zoning of the garnet and the regular decrease in KD = (Fe^{2+}/ $\text{Mg})_{\text{Grt}}/(\text{Fe}^{2+}/\text{Mg})_{\text{Cld}}$ from the garnet core to its rim, representing a temperature rise during garnet crystallization. Data suggest that lawsonite was stable at the peak UHP stage, which is supported by the presence of PMAs and quartz pseudomorphs after coesite at garnet rims. The decrease in modal amount of clinozoisite 1-bearing PMAs from garnet core to its rim show that lawsonite was not entirely used up by reaction (1.67) and some of them were stable at peak UHP conditions. The chloritoid + lawsonite assemblage outlines the highest temperature for the studied samples; i.e. peak P−T conditions of P = 27−32 kbar and T = 530−580°C for the talc-garnet-chloritoid schist agree well with the earlier estimated conditions of P = 25−28 kbar and T = 550−600°C. Minor retrograde textures comprise the growth of paragonite and chlorite replacing glaucophane and chloritoid, respectively. The existence of lawsonite and the obtained path from the prograde to peak UHP stages (27−32 kbar) under low-temperature (530−580°C) conditions suggests that talc-garnet-chloritoid schists from the Makbal Complex (480−509 Ma) underwent cold subduction to great depth (>100 km) along a low geothermal gradient (<6°C/km) during the Early Paleozoic. Dehydration reactions of hydrous minerals are supposed to occur during heating associated with subduction of rocks and the released fluids can possibly cause partial melting of the mantle wedge and/or intermediate−deep earthquakes. Orozbaev et al. (2015) established lawsonite breakdown to PMAs made up of clinozoisite 1 + kyanite + quartz ± paragonite ± phengite together with H_2O release (dehydration reaction) during isothermal decompression with the temperature being constant or slightly lowered, i.e., rapid exhumation.

References

Agard, P., Jullien, M., Goffe, B., Baronnet, A., Bouybaouene, M., 1999. TEM evidence for high-temperature (300°C) smectite in multistage clay-mineral pseudomorphs in pelitic rocks (Rif, Morocco). Eur. J. Mineral. 11, 655−668.

Al-Shanti, A.M.S., Abdel-Monem, A.A., Marzouki, F.H., 1984. Geochemistry, petrology and rubidium-strontium dating of trondhjemite and granophyre associated with Jabal Tays Ophiolite, Idsas area, Saudi Arabia. Precambrian Res. 24, 321−334.

Alvarez-Lloret, P., Rodriguez-Navarro, A.B., Falini, G., Fermani, S., Ortega-Huertas, M., 2010. Crystallographic control of the hydrothermal conversion of calcitic sea urchin spine (*Paracentrotus lividus*) into apatite. Cryst. Growth Des. 10, 5227−5232.

Åmli, R., 1975. Mineralogy and rare earth geochemistry of apatite and xenotime from the Gloserheia Granite Pegmatite, Froland, southern Norway. Am. Mineral. 60, 607−620.

Anand, R.R., Gilkes, R.J., 1984. The retention of elements in mineral pseudomorphs in lateritic saprolite from granite − a weathering budget. Aust. J. Soil Res. 22, 273−282.

Armbruster, T., 1981. On the origin of sagenites − structural coherency of rutile with hematite and spinel structure types. Neues Jahrb. Miner. 7, 328−334.

Asta, M.P., Cama, J., Ayora, C., Aceto, P., deGiudici, G., 2010. Arsenopyrite dissolution rates in O_2-bearing solutions. Chem. Geol. 273, 272−285.

Austrheim, H., Prestvik, T., 2008. Rodingitization and hydration of the oceanic lithosphere as developed in the Leka ophiolite, north-central Norway. Lithos 104, 177−198.

Barnicoat, A.C., Cartwright, I., 1997. The gabbro-eclogite transformation: an oxygen isotope and petrographic study of west Alpine ophiolites. J. Metamorph. Geol. 15, 93−104.

Basu, A., Schreiber, M.E., 2013. Arsenic release from arsenopyrite weathering: insights from sequential extraction and microscopic studies. J. Hazard. Mater. 262, 896−904.

Bates, R.L., Jackson, J.A., 1980. Glossary of Geology. American Geological Institute, Falls Church, VA.

Behr, H.J., Ahrendt, H., Martin, H., Porada, H., Rtihrs, J., Weber, K., 1983. Sedimentology and mineralogy of Upper Proterozoic playa-lake deposits in the Damara Orogen. In: Martin, H., Eder, F.W. (Eds.), Intracontinental Fold Belts: Case Studies in the Variscan Belt of Europe and the Damara Belt of Namibia. Springer, Berlin, pp. 577−610.

Bermanec, V., Horvat, M., Zigovecki Gobac, Z., Zebec, V., Scholz, R., Skoda, R., Wegner, R., Barreto, S.D.B., Dodony, I., 2012. Pseudomorphs of low microcline after adularia fourlings from the Alto da Cabeca (Boqueirao) and Morro Redondo pegmatites, Brazil. Can. Mineral. 50, 975−987.

Bischof, J.L., 1968. Catalysis, inhibition, and the calcite-aragonite problem. I. The vaterite-aragonite transformation. Amer. J. Sci. 266, 65–79.

Bischof, J.L., Fyfe, W.S., 1968. Catalysis, inhibition, and the calcite-aragonite problem. II. The aragonite-calcite transformation. Amer. J. Sci. 266, 80–90.

Blank, R.R., Fosberg, M.A., 1991. Duripans of the Owyhee Plateau region of Idaho: genesis of opal and sepiolite. Soil Sci. 152, 116–133.

Blum, J.R., 1847. Nachtrag zu den Pseudomorphosen des Mineralreich. E. Schweizerbart'sche Verlagshandlung, Stuttgart, 213 p.

Blum, J.R., 1852. Zweiter Nachtrag zu den pseudomorphosen des Mineralreichs. Julius Groos, Heidelberg, 140 p.

Blum, J.R., 1863. Dritter Nachtrag zu den Pseudomorphosen des Mineralreichs. Verlag von Ferdinand Enke, Erlangen.

Blum, J.R., 1879. Vierter Nachtrag zu den Pseudomorphosen des Mineralreichs. Carl Winter's Universitätbuchhandlung, Heidelberg, 212 p.

Blum, J.R., 1843. Die Pseudomorphosen des Mineralreichs. E. Schweizerbart'-sche Verlagshandlung, Stuttgart.

Boettcher, A.L., Wyllie, P.J., 1967. Biaxial calcite inverted from aragonite. Am. Mineral. 52, 1527–1529.

Borg, S., Liu, W., Pearce, M., Cleverley, J., MacRae, C., 2014. Complex mineral zoning patterns caused by ultra-local equilibrium at reaction interfaces. Geology 42, 415–418.

Boudreau, A.E., McCallum, I.S., 1990. Low-temperature alteration of REE-rich chlorapatite from the Stillwater Complex, Montana. Am. Mineral. 75, 687–693.

Bradley, W.H., Eugster, H.P., 1969. Geochemistry and Paleolimnology of the Trona Deposits and Associated Authigenic Minerals of the Green River Formation of Wyoming.

Brady, J.B., Markley, M.J., Schumacher, J.C., Cheneyd, J.T., Bianciardi, G.A., 2004. Aragonite pseudomorphs in high-pressure marbles of Syros, Greece. J. Struct. Geol. 26, 3–9.

Brady, K.S., Bigham, J.M., Jaynes, W.F., Logan, T.J., 1986. Influence of sulfate on Fe-oxide formation: comparisons with a stream receiving acid mine drainage. Clays Clay Miner. 34, 266–274.

Brantley, S.L., Evans, B., Hickman, S.H., Crerar, D.A., 1990. Healing of microcracks in quartz: implications for fluid flow. Geology 18, 136–139.

Breithaupt, A., 1815. Über die Aechtheit der Krystalle. Craz und Gerlack, Freiberg, 63 p.

Brown, A.C., 2005. Refinements for footwall red-bed diagenesis in the sediment-hosted stratiform copper deposits model. Econ. Geol. 100, 765–771.

Brown, W.H., Fyfe, W.S., Turner, F.J., 1962. Aragonite in California glaucophane schists, and the kinetics of the aragonite–calcite transformation. J. Petrol. 3, 566–582.

Bruno, M., Compagnoni, R., Hirajima, T., Rubbo, M., 2002. Jadeite with the Ca-Eskola molecule from an ultra-high pressure metagranodiorite, Dora-Maira Massif, Western Alps. Contrib. Mineral. Petrol. 142, 515–519.

Bruno, M., Compagnoni, R., Rubbo, M., 2001. The ultra-high pressure coronitic and pseudomorphous reactions in a metagranodiorite from the Brossasco-Isasca Unit, Dora-Maira Massif, western Italian Alps: a petrographic study and equilibrium thermodynamic modeling. J. Metamorph. Geol. 19, 33–43.

Brush, G.J., Dana, E.S., 1879. On the mineral ocality in Fairfield County, Connecticut. Am. J. Sci. l17, 359–368. Second paper.

Brush, G.J., Dana, E.S., 1880. On the mineral locality at Branchville, Connecticut. Fourth paper. Spodumene and the results of its alteration. Am. J. Sci. l18, 257–285.

Burt, D.M., 1989. Compositional and phase relations among rare earth element minerals. In: Lipin, B.R., McKay, G.A. (Eds.), Geochemistry and Mineralogy of Rare Earth Elements, vol. 21. Mineralogical Society of America, Washington, DC, pp. 259–307.

Caillaud, J., Proust, D., Righi, D., 2006. Weathering sequences of rock-forming minerals in a serpentinite: influence of microsystems on clay mineralogy. Clays Clay Miner 54, 87–100.

Carlson, W.D., Rosenfeld, J.L., 1981. Optical determination of topotactic aragonite–calcite growth kinetics: metamorphic implications. J. Geol. 89, 615–638.

Carr, P.F., Perkins, M., Moradian, A., 1996. Large pseudomorphous trapezohedra of analcime and pumpellyite after leucite, Aghda area, Central Iran. Mineral. Petrol. 58, 23–32.

Cassedanne, J.P., Menezes, S.D.O., 1989. "Pseudoleucite" pseudomorphs from Rio Das Ostras, Brazil. Mineral. Rec. 20, 439–440.

Centrella, S., Austrheim, H., Putnis, A., 2015. Coupled mass transfer through a fluid phase and volume preservation during the hydration of granulite: an example from the Bergen Arcs, Norway. Lithos 236–237, 245–255.

Černý, P., 1968. Beryllumwandlungen in Pegmatiten—Verlauf und Produkte. N. Jb. Miner. Ahb. 108, 166–180.

Černý, P., 1978. Alteration of pollucite in some pegmatites of southeastern Manitoba. Can. Mineral 16, 89–95.

Cesare, B., 1999. Multi-stage pseudomorphic replacement of garnet during polymetamorphism: 1. Microstructures and their interpretation. J. Metamorph. Geol. 17, 723–734.

Chakhmouradian, A.R., Mitchell, R.H., 1997. Compositional variarion of perovskite-group minerals from the carbonatite complexes of the Kola alkaline Province, Russia. Can. Mineral. 35, 1293–1310.

Chakhmouradian, A.R., Mitchell, R.H., 1998. Compositional variation of perovskite-group minerals from the Khibina complex, Kola Peninsula. Can. Mineral. 36, 953–969.

Chatterjee, N.D., 1971. Preliminary results on the synthesis and upper stability of margarite. Naturwissenschaften 58, 147.

Chatterjee, N.D., 1972. The upper stability limit of the assemblage paragonite + quartz and its natural occurrence. Contrib. Mineral. Petrol 34, 288–303.

Chatterjee, N.D., 1976. Margarite stability and compatibility relations in the system $CaO\text{-}Al_2O_3\text{-}SiO_2-H_2O$ as a pressure–temperature indicator. Am. Mineral 61, 699–709.

Chukhrov, F.V., Zvyagin, B.B., Gorshkov, A.I., Yermilova, L.P., Kovovushkin, V.V., Rudnitskaya, S.Y., Yabukovskaya, N.Y., 1976. Feroxyhyte, a new modification of FeOOH: SSSR Izv. Ser. Geol. 5, 15–24. Transl.: Int. Geol. Rev., V, 19: 873–889.

Cissarz, A., Jones, W.R., 1931. German-English Geological Terminology. Thomas Murby and Co., London.

Cole, D.R., 2000. Isotopic exchange in mineral-fluid systems. IV. The crystal chemical controls on oxygen isotope exchange rates in carbonate-H_2O and layer silicate-H_2O systems. Geochim. Cosmochim. Acta 64, 921–931.

Cole, D.R., Chakraborty, S., 2001. Rates and mechanisms of isotopic exchange. In: Valley, J.W., Cole, D. (Eds.), Stable Isotope Geochemistry, vol. 43. Mineralogical Society of America, Chantilly, VA, pp. 83–223.

Cole, D.R., Larson, P.B., Riciputi, L.R., Mora, C.I., 2004. Oxygen isotope zoning profiles in hydrothermally altered feldspars: estimating the duration of water—rock interaction. Geology 32, 29—32.

Courtin-Nomade, A., Bril, H., Neel, C., Lenain, J.-F., 2003. Arsenic in iron cements developed within tailings of a former metalliferous mine—Enguialès, Aveyron, France. Appl. Geochem. 18, 395—408.

Craig, J.R., Vaughan, D.J., Higgins, J.B., 1979. Phase relations in the Cu—Co—S system and mineral associations of the carrolite (CuCo$_2$S$_4$)—linnaeite (Co$_3$S$_4$) series. Econ. Geol. 74, 657—671.

Daczko, N.R., Clarke, G.L., Klepeis, K.A., 2002. Kyanite-paragonite-bearing assemblages, northern Fiordland, New Zealand: rapid cooling of the lower crustal root to a Cretaceous magmatic arc. J. Metamorph. Geol. 20, 887—902.

Dana, J.D., 1837. A System of Mineralogy. George P. Putnam & Co., New York.

Dana, J.D., 1844. A System of Mineralogy. George P. Putnam & Co., New York.

Dana, J.D., 1845. Observations on pseudomorphism. Amer. J. Sci. Ser. 1, 81—92.

Dana, J.D., 1850. A System of Mineralogy. George P. Putnam & Co., New York.

Dana, J.D., 1854. A System of Mineralogy. George P. Putnam & Co., New York.

Daval, D., Martinez, I., Guigner, J.-M., Hellmann, R., Corvisier, J., Findling, N., Dominici, C., Goffé, B., Guyot, F., 2009. Mechanism of wollastonite carbonation deduced from micro- to nanometer length scale observations. Am. Mineral. 94.

Davies, G.R., Nixon, P.H., Pearson, D.G., Obata, M., 1993. Tectonic implications of graphitized diamonds from the Ronda peridotite massif, southern Spain. Geology 21, 471—474.

Delaney, J.M., Helgeson, H.C., 1978. Calculation of the thermodynamic consequence of dehydration in subducting oceanic crust to 100 kb and 600°C. Am. J. Sci 278, 638—686.

De Lurio, J.L., Frakes, L.A., 1999. Glendonites as a paleoenvironmental tool: implications for early cretaceous high latitude climates in Australia. Geochim. Cosmochim. Acta 63, 1039—1048.

Delvigne, J., 1965. Pédogenèse en zone tropicale. La formation des minéraux secondaires en milieu ferralitique. Mémoires vol. 13. O.R.S.T.O.M., Paris, 177 p.

Delvigne, J., Bisdom, E.B.A., Sleeman, J., Stoops, G., 1979. Olivines, their pseudomorphs and secondary products. Pedologie 29, 247—309.

Dempster, T., Jess, S.A., 2015. Ikaite pseudomorphs in Neoproterozoic Dalradian slates record Earth's coldest metamorphism. J. Geol. Soc. 172, 459—464.

Dempster, T.J., Persano, C., 2006. Low temperature thermochronology: resolving geotherm shapes or denudation histories? Geology 34, 73—76.

Dousma, J., den Ottelander, D., de Bruyn, P.L., 1979. The influence of sulfate ions on the formation of iron(III) oxides. J. Inorg. Nucl. Chem. 41, 1565—1568.

Drits, V.A., Ivanovskaya, T.A., Sakharov, B.A., Gor'kova, N.V., Karpova, G.V., Pokrovskaya, E.V., 2001. Pseudomorphous replacement of globular glauconite by mixed-layer chlorite-berthierine in the outer contact of dike: evidence from the Lower Riphean Ust'-Il'ya formation, Anabar uplift. Lithol. Miner. Resour. 36, 337—352 (Transl. of Litol. Polezn. Iskop.).

Dungan, M.A., 1979. A microprobe study of antigorite and some serpentine pseudomorphs. Can. Mineral. 17, 771—784.

Dunn, J.G., Howes, V.L., 1996. The oxidation of violarite. Thermochim. Acta 305—316, 282/283.

Dutrizac, J.E., 1983. Factors affecting alkali jarosite precipitation. Metal. Mat. Trans. B 14, 531—539.

Dutrow, B.L., Foster Jr., C.T., Henry, D.J., 1999. Tourmaline-rich pseudomorphs in sillimanite zone metapelites: demarcation of an infiltrate. Am. Mineral. 84, 794—805.

Dutrow, B.L., Foster Jr., C.T., Whittington, J., 2008. Prograde muscovite-rich pseudomorphs as indicators of conditions during metamorphism: an example from NW maine. Am. Mineral. 93, 300—314.

Dutrow, B.L., Holdaway, M.J., 1989. Experimental determination of the upper thermal stability of Fe-staurolite þ quartz at medium pressures. J. Petrol. 30, 229—248.

Edelman, S.H., Day, H.W., Moores, E.M., Zigan, S.M., Murphy, T.P., Hacker, B.R., 1989. Structure across a Mesozoic ocean-continent suture zone in the northern Sierra Nevada, California. Geol. Soc. Am. Spec. Pap. 224, 56.

El Atrassi, F., Brunet, F., Bouybaouène, M., Chopin, C., Chazot, G., 2011. Melting textures and microdiamonds preserved in graphite pseudomorphs from the Beni Bousera peridotite massif, Morocco. Eur. J. Mineral. 23, 157—168.

El Atrassi, F., Brunet, F., Chazot, G., Bouybaouene, M., Chopin, C., 2013. Metamorphic and magmatic overprint of garnet pyroxenites from the Beni Bousera massif (northern Morocco): petrography, mineral chemistry and thermobarometry. Lithos 179, 231—248.

El Tabakh, M., Grey, K., Pirajno, F., Schreiber, B.C., 1999. Pseudomorphs after evaporitic minerals interbedded with 2.2 Ga stromatolites of the Yerrida basin, Western Australia: origin and significance. Geology 27, 871—874.

Elton, N.J., Hooper, J.J., Holyer, V.A.D., 1997. An occurrence of stevensite and kerolite in the Devonian Crousa gabbro at Dean Quarry, The Lizard, Cornwall, England. Clay Miner. 32, 241—252.

Engvik, A.K., Golla-Schindler, U., Berndt, J., Austrheim, H., Putnis, A., 2009. Intragranular replacement of chlorapatite by hydroxy-fluor-apatite during metasomatism. Lithos 112, 236—246.

Engvik, A.K., Putnis, A., Fitz Gerald, J.D., Austrheim, H., 2008. Albitization of granitic rocks: the mechanism of replacement of oligoclase by albite. Can. Mineral. 46, 1401—1415.

Estner, F.J.A., 1794—1804. Versuch einer Mineralogie für Anfänger und Liebhaber nach des Herrn Bergcommissionsraths Werner's Methode. Jos. Georg Oehler, Wien.

Estrade, G., Salvi, S., Beziat, D., Rakotovao, S., Rakotondrazafy, R., 2014. REE and HFSE mineralization in peralkaline granites of the Ambohimirahavavy alkaline complex, Ampasindava peninsula, Madagascar. J. Afr. Earth Sci. 94, 141—155.

Etschmann, B., Brugger, J., Pearce, M.A., Ta, C., Brautigan, D., Jung, M., Pring, A., 2014. Grain boundaries as microreactors during reactive fluid flow: experimental dolomitization of a calcite marble. Contrib. Mineral. Petrol. 168, 1—12.

Eugster, H.P., 1969. Inorganic bedded cherts from the Magadi area, Kenya. Contrib. Mineral. Petrol. 22, 1—31.

Eugster, H.P., 1970. Chemistry and origin of the brines of Lake Magadi, Kenya. Mineral. Soc. Am. Spec. Publ. 3, 215—235.

Eysel, W., Roy, D.M., 1975. Topotactic reaction of aragonite to hydroxyapatite. Z. Kristallogr. 141, 11—24.

Fagan, T.J., Day, H.W., 1994. Metamorphism of the Slate Creek complex, Central Belt, northern Sierra Nevada: pre-orogenic volcanic setting? Geol. Soc. Am. Abstr. Programs 26, 50.

Fagan, T.J., Day, H.W., 1997. Formation of amphibole after clinopyroxene by dehydration reactions. Implications for pseudomorphic replacement and mass fluxes. Geology 25, 395–398.

Feng, Y., Samson, I.M., 2015. Replacement processes involving high field strength elements in the T zone, Thor Lake rare-metal deposit. Can. Mineral. 53, 31–60.

Fernandez-Diaz, L., Pina, C.M., Astilleros, J.M., Sanchez-Pastor, N., 2009. The carbonatation of gypsum: pathways and pseudomorph formation. Am. Mineral. 94, 1223–1234.

Filippi, M., 2004. Oxidation of the arsenic-rich concentrate at the Přebuz abandoned mine (Erzgebirge Mts, CZ): mineralogical evolution. Sci. Total Environ. 322, 271–282.

Filippi, M., Drahota, P., Machovic, V., Bohmova, V., Mihaljevic, M., 2015. Arsenic mineralogy and mobility in the arsenic-rich historical mine waste dump. Sci. Total Environ. 536, 713–728.

Fleet, M.E., Liu, X., Pan, Y., 2000. Rare-earth elements in chlorapatite [$Ca_{10}(PO_4)_6Cl_2$]: uptake, site preference and degradation of monoclinic structure. Am. Mineral. 85, 1437–1446.

Fleet, M.E., Pan, Y., 1995. Site preference of rare earth elements in fluorapatite. Am. Mineral. 80, 329–335.

Fleet, M.E., Pan, Y., 1997. Site preference of rare earth elements in fluorapatite: binary (LREE + HREE)-substituted crystals. Am. Mineral. 82, 870–877.

Folk, R.L., Assereto, R., 1976. Comparitive fabrics of length-slow and length-fast calcite and calcitised aragonite in a Holocene speleothem, Carlsbad Caverns, New Mexico. J. Sediment. Petrol. 46, 486–496.

Foster Jr., C.T., 1983. Thermodynamic models of biotite pseudomorphs after staurolite. Am. Mineral. 68, 389–397.

Foster Jr., C.T., 1981. A thermodynamic model of mineral segregations in the lower sillimanite zone near Rangeley, Maine. Am. Mineral. 66, 260–277.

Francis, C.A., Lange, D.E., Pitman, L.C., Croft, W.J., Lillie, R.C., 1997. Barite after paralstonite, a new pseudomorph from Cave-in-Rock, Illinois. Mineral. Rec. 28, 443–446.

Frank, J.R., Carpenter, A.B., Oglesby, T.W., 1982. Cathodoluminescence and composition of calcite cement in the Taum Sauk limestone (Upper Cambrian), southeast Missouri. J. Sediment. Petrol. 52, 631–638.

Franz, G., Althaus, E., 1977. The stability relations of the paragenesis paragonite-zoisite-quartz. Neues Jahrb. Mineral. Abh 130, 159–167.

Freyssinet, P., Butt, C., Morris, R., Piantone, P., 2005. Ore-forming processes related to lateritic weathering. Econ. Geol. 681–722, 100th Anniv. Vol.

Frisch, W., 1975. Die Wolfram-Lagerstätte Gifurwe (Rwanda) und die Genese der zentralafrikanischen Reinit-Lagerstätten [in German]. Jahrb. Geol. Bundesanst. 118, 119–191.

Frisia, S., Borsato, A., Fairchild, I.J., McDermott, F., 2000. Calcite fabrics, growth mechanisms, and environments of formation in speleothems from the Italian Alps and Southwestern Ireland. J. Sediment. Res. 70, 1183–1196.

Frisia, S., Borsato, A., Fairchild, I.J., McDermott, F., Selmo, E.M., 2002. Aragonite-calcite relationships in speleothems (Grotte de Clamouse, France): environment, fabrics, and carbonate geochemistry. J. Sediment. Res. 72, 687–699.

Frondel, C., 1935. Catalogue of minernal pseudomorphs in the American Museum. Bull. Amer. Mus. Natur. Hist. LXVII, 389–426.

Galuskin, E., Galuskina, I., Winiarska, A., 1995. Epitaxy of achtarandite on grossular – the key to the problem of achtarandite. Neues Jahrb. Mineral. Monatsh. 306–320.

Galuskina, I., Galuskin, E., Sitarz, M., 1998. Atoll hydrogarnets and mechanism of the formation of achtarandite pseudomorphs. Neues Jahrb. Mineral. Monatsh 49–62.

Galuskina, I., Galuskin, E., Sitarz, M., 2001. Evolution of morphology and composition of hibschite, Wiluy River, Yakutia. Neues Jahrb. Mineral. Monatsh 49–66.

Geisler, T., Berndt, J., Meyer, H.W., Pollok, K., Putnis, A., 2004. Low-temperature aqueous alteration of crystalline pyrochlore: correspondence between nature and experiment. Mineral. Mag. 68, 905–922.

Geisler, T., Poeml, P., Stephan, T., Janssen, A., Putnis, A., 2005a. Experimental observation of an interface-controlled pseudomorphic replacement reaction in a natural crystalline pyrochlore. Am. Mineral. 90, 1683–1687.

Geisler, T., Schaltegger, U., Tomaschek, F., 2007. Re-equilibration of zircon in aqueous fluids and melts. Elements 3, 43–50.

Geisler, T., Seydoux-Guillaume, A.M., Poeml, P., Golla-Schindler, U., Berndt, J., Wirth, R., Pollok, K., Janssen, A., Putnis, A., 2005b. Experimental hydrothermal alteration of crystalline and radiation-damaged pyrochlore. J. Nucl. Mater. 344, 17–23.

Gillet, P., Gérard, Y., Willaime, C., 1987. The calcite–aragonite transition: mechanism and microstructures induced by the transformation stresses and strain. Bull. Mineral. 110, 481–496.

Gleissner, P., Glodny, J., Franz, G., 2007. Rb-Sr isotopic dating of pseudomorphs after lawsonite in metabasalts from the Glockner nappe, Tauern Window, Eastern Alps. Eur. J. Mineral. 19, 723–734.

Glikin, A.E., Leontieva, O.A., Sinai, M.Y., 1994. Mechanisms of exchange of isomorphous components between crystal and solution and macrodefectiveness in secondary crystals. J. Struct. Chem. 35.

Godard, G., 2009. Two orogenic cycles recorded in eclogite-facies gneiss from the southern Armorican Massif (France). Eur. J. Mineral. 21, 1173–1190.

Golden, D.C., Chen, C.C., Dixon, J.B., Tokashiki, Y., 1988. Pseudomorphic replacement of manganese oxides by iron oxide minerals. Geoderma 42, 199–211.

Golden, D.C., Ming, D.W., Morris, R.V., Graff, T.G., 2008. Hydrothermal synthesis of hematite spherules and jarosite: implications for diagenesis and hematite spherule formation in sulfate outcrops at Meridiani Planum, Mars. Am. Mineral. 93, 1201–1214.

Goldmann, S., Melcher, F., Gaebler, H.-E., Dewaele, S., De Clercq, F., Muchez, P., 2013. Mineralogy and trace element chemistry of ferberite/reinite from tungsten deposits in central Rwanda. Minerals 3, 121–144.

Golightly, J., 1981. Nickeliferous laterite deposits. Econ. Geol. 75, 710–735.

Golightly, J.P., Arancibia, O.N., 1979. The chemical composition and infrared spectrum of nickel- and iron-substituted serpentine from a nickeliferous laterite profile, Soroako, Indonesia. Can. Mineral. 17, 719–728.

Gomez-Pugnaire, M.T., Franz, G., Lopez Sanchez-Vizcaino, V., 1994. Retrograde formation of NaCl-scapolite in high pressure metaevaporites from the Cordilleras Beticas (Spain). Contrib. Mineral. Petrol. 116, 448–461.

Gomez-Pugnaire, M.T., Visona, D., Franz, G., 1985. Kyanite, margarite and paragonite in pseudomorphs in amphibolitized eclogites from the Betic Cordilleras, Spain. Chem. Geol. 50, 129−141.

Grant, J.A., 1986. The isocon diagram; a simple solution to Gresens' equation for metasomatic alteration. Econ. Geol. 81, 1976−1982.

Grant, J.A., 2005. Isocon analysis: a brief review of the method and applications. Phys. Chem. Earth A/B/C 30, 997−1004.

Grapes, R., Li, X.-P., 2010. Disequilibrium thermal breakdown of staurolite: a natural example. Eur. J. Mineral. 22, 147−157.

Gratacap, L.P., 1912. A Popular Guide to Minerals. D. Van Nostrand Company, New York.

Grguric, B.A., 2002. Hypogene violarite of exsolution origin from Mount Keith, Western Australia: field evidence for a stable pentlandite-violarite tie line. Mineral. Mag. 66, 313−326.

Guidotti, C.V., 1968. Prograde muscovite pseudomorphs after staurolite in the Rangeley-Oquossoc areas, Maine. Am. Mineral. 53, 1368−1376.

Gurney, J.J., 1989. Diamonds. In: Ross, J. (Ed.), Kimberlites and Related Rocks, Proceedings of the 4th International Kimberlite Conference, Volume 2, Section 935-965, Geological Society of Australia Special Publication, Perth, Vol. 14, pp. 935−965.

Güven, N., Carney, L.L., 1979. The hydrothermal transformation of sepiolite to stevensite and the effect of added chlorides and hydroxides. Clays Clay Miner. 27, 253−260.

Haidinger, W., 1827. On the gradual changes which take place in the interior of cupriferous minerals, while their external form remains the same. Edinburgh J. Sci. 7, 126−133.

Haidinger, W., 1828. On the parasitic formation of mineral species, depending upon gradual changes which take place in the interior of minerals, while their external form remains the same. Edinburgh J. Sci. 9, 275−292.

Haidinger, W., 1829. On the parasitic formation of mineral species, depending upon gradual changes which take place in the interior of minerals, while their external form remains the same. Edinburgh J. Sci. 10, 86−95.

Harlov, D.E., Andersson, U.B., Förster, H.-J., Nyström, J.O., Dulski, P., Broman, C., 2002a. Apatite-monazite relations in the Kiirunavaara magnetite-apatite ore, northern Sweden. Chem. Geol. 191.

Harlov, D.E., Förster, H.-J., 2002. High-grade fluid metasomatism on both a local and regional scale: the Seward Peninsula, Alaska and the Val Strona di Omegna, Ivrea-Verbano zone, northern Italy. Part II: phosphate mineral chemistry. J. Petrol. 43, 801−824.

Harlov, D.E., Förster, H.-J., 2003. Fluid-induced nucleation of (Y+REE)-phosphate minerals within apatite: nature and experiment. Part II. Fluorapatite. Am. Mineral. 88, 1209−1229.

Harlov, D.E., Förster, H.-J., Nijland, T.G., 2002b. Fluid-induced nucleation of (Y + REE)-phosphate minerals within apatite: nature and experiment. Part I. Chlorapatite. Am. Mineral. 87, 245−261.

Harlov, D.E., Förster, H.J., Schmidt, C., 2003. High P-T experimental metasomatism of a fluorapatite with significant britholite and fluorellestadite components: implications for LREE mobility during granulite-facies metamorphism. Mineral. Mag. 67, 61−72.

Harlov, D.E., Wirth, R., Förster, H.-J., 2005. An experimental study of dissolution−reprecipitation in fluorapatite: fluid infiltration and the formation of monazite. Contrib. Mineral. Petrol. 150, 268−286.

Harlov, D.E., Wirth, R., Hetherington, C.J., 2007. The relative stability of monazite and huttonite at 300−900°C and 200−1000 MPa: metasomatism and the propagation of metastablemineral phases. Am. Mineral. 92, 1652−1664.

Haüy, R.J., 1801. Traité de Minéralogie. Chez Louis, Paris.

Heinrich, W., Althaus, E., 1980. Die obere Stabilitätsgrenze von Lawsonit plus Albit bzw. Jadeit. Fortschr. Mineral. Beiheft 58, 49−50.

Hellmann, R., Penisson, J.M., Hervig, R.L., Thomassin, J., Abrioux, M.F., 2003. An EFTEM/HRTEM high-resolution study of the near surface of labradorite feldspar altered at acid pH: evidence for interfacial dissolution−reprecipitation. Phys. Chem. Miner. 30, 192−197.

Henry, D.J., Dutrow, B.L., Selverstone, J., 2002. Compositional polarity in replacement tourmaline—an example from the Tauern Window, Eastern Alps. Geol. Mater. Res. 4, 23 p.

Hiroi, Y., Hokada, T., Beppu, M., Motoyoshi, Y., Shimura, T., Yuhara, M., Shiraishi, K., Grantham, G.H., Knoper, M.W., 2001. New evidence for prograde metamorphism and partial melting of Mg-Al-rich granulites from western Namaqualand, South Africa. Mem. Natl. Inst. Polar Res. Spec. Issue 55, 87−104.

Hoffmann, C.A.S., 1811−1817. Handbuch der Mineralogie. Freiberg, Graß und Gerlach.

Holland, T.J.B., 1979. High water activities in the generation of the high pressure kyanite eclogites in the Tauern Window, Austria. J. Geol 87, 1−27.

Hövelmann, J., Putnis, A., Geisler, T., Schmidt, B.C., Golla-Schindler, U., 2010. The replacement of plagioclase feldspars by albite: observations from hydrothermal experiments. Contrib. Mineral. Petrol. 159, 43−59.

Hughes, J.M., Cameron, M., Mariano, A.N., 1991. Rare-earth-element ordering and structural variations in natural rare-earth-bearing apatites. Am. Mineral. 76, 1165−1173.

Jamtveit, B., Putnis, C.V., Malthe-Sorenssen, A., 2009. Reaction induced fracturing during replacement processes. Contrib. Mineral. Petrol. 157, 127−133.

Janssen, A., Putnis, A., Geisler, T., Putnis, C.V., 2008. The mechanism of experimental oxidation and leaching of ilmenite in acid solution. Publ. Australas. Inst. Min. Metall. 8, 503−506.

Janssen, A., Putnis, A., Geisler, T., Putnis, C.V., 2010. The experimental replacement of ilmenite by rutile in HCl solutions. Mineral. Mag. 74, 633−644.

Jeong, G.Y., Lee, B.Y., 2003. Secondary mineralogy and microtextures of weathered sulfides and manganoan carbonates in mine waste-rock dumps, with implications for heavy-metal fixation. Am. Mineral. 88, 1933−1942.

Johnston, J.D., 1995. Pseudomorphs after ikaite in a glaciomarine sequence in the Dalradian of Donegal, Ireland. Scottish J. Geology 31, 29−49.

Johs, M., 1981. Pseudomorphososen ein kurzer geschichtlicher Überblick. Lapis 6, 36−37.

Julien, A.A., 1879. On spodumene and its alrerarions from granite-veinso of Hampshire County, Massachusetts. Ann. N. Y. Acad. Sci. l, 318−359.

Kasioptas, A., Geisler, T., Perdikouri, C., Trepmann, C., Gussone, N., Putnis, A., 2011. Polycrystalline apatite synthesized by hydrothermal replacement of calcium carbonates. Geochim. Cosmochim. Acta 75, 3486−3500.

Kasioptas, A., Geisler, T., Putnis, C.V., Perdikouri, C., Putnis, A., 2010. Crystal growth of apatite by replacement of an aragonite precursor. J. Cryst. Growth 312, 2431−2440.

Kasioptas, A., Perdikouri, C., Putnis, C.V., Putnis, A., 2008. Pseudomorphic replacement of single calcium carbonate crystals by polycrystalline apatite. Mineral. Mag. 72, 77−80.

Kendall, A.C., Tucker, M.E., 1973. Radiaxial fibrous calcite: a replacement after acicular carbonate. Sedimentology 20, 365−389.

King, H.E., Pluemper, O., Geisler, T., Putnis, A., 2011. Experimental investigations into the silicification of olivine: implications for the reaction mechanism and acid neutralization. Am. Mineral. 96, 1503–1511.

Klasa, J., Ruiz-Agudo, E., Wang, L.J., Putnis, C.V., Valsami-Jones, E., Menneken, M., Putnis, A., 2013. An atomic force microscopy study of the dissolution of calcite in the presence of phosphate ions. Geochim. Cosmochim. Acta 117, 115–128.

Kocks, U.F., Tome, C.N., Wenk, H.R., 2001. Texture and Anisotropy. Preferred Orientations in Polycrystals and their Effect on Materials Properties. Cambridge University Press, Cambridge, UK.

Kulke, H., 1978. Tektonik und Petrographic einer Salinarformation am Beispiel der Trias des Atlassystems (NW-Afrika). Geotektonische Forsch 55, 1–158.

Kusebauch, C., John, T., Whitehouse, M.J., Klemme, S., Putnis, A., 2015. Distribution of halogens between fluid and apatite during fluid-mediated replacement processes. Geochim. Cosmochim. Acta 170, 225–246.

Labotka, T.C., Cole, D.R., Fayek, M., Riciputi, L.R., Stadermann, F.J., 2004. Coupled cation and oxygen-isotope exchange between alkali feldspar and aqueous chloride solution. Am. Mineral. 89, 1822–1825.

Lahti, S.I., 1981. On the Granitic Pegmatites of the Eräjärvi Area in Southern Finland.

Lahti, S.I., 1988. Occurrence and mineralogy of the margarite- and muscovite-bearing pseudomorphs after topaz in the Juurakko pegmatite, Orivesi, southern Finland. Bull. Geol. Soc. Finl. 60, 27–43.

Lahti, S.I., Saikkonen, R., 1985. Bityite 2M1 from Eräjärvi compared with related lithium-beryllium brittle micas. Bull. Geol. Soc. Finl. 57, 207–215.

Lambrecht, G., Diamond, L.W., 2014. Morphological ripening of fluid inclusions and coupled zone-refining in quartz crystals revealed by cathodoluminescence imaging: implications for CL-petrography, fluid inclusion analysis and trace-element geothermometry. Geochim. Cosmochim. Acta 141, 381–406.

Landgrebe, G., 1841. Über die Pseudomorphosen im Mineralreiche und verwandte Erscheinungen. Verlag J.J. Bohné, Cassel.

Landmesser, M., 1994. Zur Entstehung von Kieselhölzern. extraLapis 7, 49–80.

Landmesser, M., 1995. Mobility by metastability: silica transport and accumulation at low temperatures. Chem. Erde 55, 149–176.

Landmesser, M., 1998. "Mobility by metastability" in sedimentary and agate petrology. Applications. Chem. Erde 58, 1–22.

Larsen, D., 1994. Origin and paleoenvironmental significance of calcite pseudomorphs after ikaite in the Oligocene Creede Formation, Colorado. J. Sediment. Res. A 64, 593–603.

Leitner, C., Neubauer, F., Marschallinger, R., Genser, J., Bernroider, M., 2013. Origin of deformed halite hopper crystals, pseudomorphic anhydrite cubes and polyhalite in Alpine evaporites (Austria, Germany). Int. J. Earth Sci. 102, 813–829.

Leoni, L., Sartori, F., 1983. Stevensite pseudomorphous after periclase. Neues Jahrb. Mineral. Monatsh. 556–562.

Lewis, J.F., Draper, G., Proenza Fernández, J.A., Espaillat, J., Jiménez, J., 2006. Ophiolite-related ultramafic rocks (Serpentinites) in the Caribbean region: a review of their occurrence, composition, origin, emplacements and Ni-laterite soil formation. Geol. Acta Mineral. Petrogr. 4, 237–263.

Lillie, R.C., 1988. Minerals of the Harris Creek fluorspar district, Hardin County, Illinois. Rocks Miner. 63, 210–226.

Llorca, S., Monchoux, P., 1991. Supergene cobalt minerals from New Caledonia. Can. Mineral. 29, 149–161.

London, D., Burt, D.M., 1982. Alteration of spodumene, montebrasite and lithiophilite in pegmatites of the White Picacho District, Arizona [USA]. Am. Mineral. 67, 97–113.

Lowenstein, T., 1982. Primary features in a potash evaporite deposit, the Permian Salado Formation of West Texas and New Mexico [USA]. SEPM Core Workshop 3, 276–304.

Majumdar, A.S., King, H.E., John, T., Kusebauch, C., Putnis, A., 2014. Pseudomorphic replacement of diopside during interaction with $(Ni,Mg)Cl_2$ aqueous solutions: implications for the Ni-enrichment mechanism in talc- and serpentine-type phases. Chem. Geol. 380, 27–40.

Majzlan, J., Drahota, P., Filippi, M., 2014a. Parageneses & crystal chemistry of As minerals. In: Bowell, R.J., Alpers, C.N., Jamieson, H.E., Nordstrom, D.K., Majzlan, J. (Eds.), Arsenic, Environmental Geochemistry, Mineralogy, and Microbiology, vol. 79. Mineralogical Society of America, Chantilly, VA, pp. 17–184.

Majzlan, J., Plášil, J., Škoda, R., Gescher, J., Kögler, F., Rusznyak, A., Küsel, K., Mangold, S., Rothe, J., 2014b. Arsenic-rich Acid mine water with extreme arsenic concentration: mineralogy, geochemistry, microbiology, and environmental implications. Environ. Sci. Technol. 2, 13685–13693.

Malthe-Sørenssen, A., Jamtveit, B., Meakin, P., 2006. Fracture patterns generated by diffusion controlled volume changing reactions. Phys. Rev. Letters 96, 245501.

Mariga, J., Ripley, E.M., Li, C., McKeegan, K.D., Schmidt, A., Groove, M., 2006. Oxygen isotopic disequilibrium in plagioclase-corundum-hercynite xenoliths from the Voisey's Bay Intrusion, Labrador, Canada. Earth Planet. Sci. Lett. 248, 263–275.

Marshall, D.J., 1988. Cathodoluminescence of Geological Materials. Unwin Hyman, Boston, Sydney, 146 p.

Martinez, F.J., Reche, J., Arboleya, M.L., Julivert, M., 2004. Retrograde replacement of andalusite by Ca-Na mica in chloritoid-bearing metapelites. PTX modelling of rocks with different Al content in the MnNCKFMASH system. J. Metamorph. Geol. 22, 777–792.

McClay, K.R., Carlile, D.G., 1978. Mid-proterozoic sulfate evaporites at Mount Isa mine, Queensland, Australia. Nature 274, 240–241.

McGuire, M.M., Jallad, K.N., Ben-Amotz, D., Hamers, R.J., 2001. Chemical mapping of elemental sulfur on pyrite and arsenopyrite surfaces using near-infrared Raman imaging microscopy. Appl. Surf. Sci. 178, 105–115.

Merino, E., Nahon, D., Wang, Y., 1993. Kinetics and mass transfer of pseudomorphic replacement: application to replacement of parent minerals and kaolinite by aluminum, iron, and manganese oxides during weathering. Am. J. Sci. 293, 135–155.

Metz, R., 1964. Pseudomorphosen-Trügerische Kristallgestalten. Aufschluss 15, 112–118.

Milke, R., Metz, P., 2002. Experimental investigation on the kinetics of the reaction wollastonite + calcite + anorthite = grossular + CO_2: implications for porphyroblast growth. Am. J. Sci. 302, 312–345.

Misra, K.C., Fleet, M.E., 1974. Chemical composition and stability of violarite. Econ. Geol. 69, 391–403.

Mitchell, R.H., Chakhmouradian, A.R., 1998. Instability of perovskite in a CO_2-rich environment: examples from carbonatite and kimberlite. Can. Mineral. 36, 939–952.

Morad, S., Marfil, R., Andres de la Pena, J., 1989. Diagenetic potassium-feldspar pseudomorphs in the Triassic Buntsandstein sandstones of the Iberian Range, Spain. Sedimentology 36, 635–650.

Motoyoshi, Y., Hensen, B.J., 1989. Sapphirine-quartz-orthopyroxene symplectites after cordierite in the Archean Napier Complex, Antarctica: evidence for a counterclockwise P-T path? Eur. J. Mineral. 1, 467–471.

Muchez, P., Vanderhaeghen, P., Desouky, H., Schneider, J., Boyce, A., Dewaele, S., Cailteux, J., 2008. Anhydrite pseudomorphs and the origin of stratiform Cu-Co ores in the Katangan Copperbelt (Democratic Republic of Congo). Miner. Deposita 43, 575–589.

Nassau, K., Gallagher, P.K., Miller, A.E., Graedel, T.E., 1987. The characterization of patina components by x-ray diffraction and evolved gas analysis. Corros. Sci. 27, 669–684.

Naumann, C.F., 1846. Elemente der Mineralogie. Verlag von Wilhelm Engelmann, Leipzig.

Navrotsky, A., 2002. Thermochemistry, energetic modelling and systematics. In: Gramaccioli, C.M. (Ed.), Energy Modelling in Minerals, vol. 4. Eötvös University Press, Budapest, pp. 5–31.

Nesbitt, H.W., Muir, I.J., 1998. Oxidation states and speciation of secondary products on pyrite and arsenopyrite reacted with mine waste waters and air. Mineral. Petrol. 62, 123–144.

Newton, R.C., Kennedy, G.C., 1963. Some equilibrium reactions in the join $CaAl_2Si_2O_8$-H_2O. J. Geophys. Res 68, 2967–2983.

Nickel, E.H., 1973. Violarite, a key mineral in the supergene alteration of nickel sulfide ores. Pap. West. Aust. Conf. Australas. Inst. Min. Metall. 111–116.

Niedermeier, D.R.D., Putnis, A., Geisler, T., Golla-Schindler, U., Putnis, C.V., 2009. The mechanism of cation and oxygen isotope exchange in alkali feldspars under hydrothermal conditions. Contrib. Mineral. Petrol. 157, 65–76.

Nishimura, Y., Satoh, H., Tsukamoto, K., Yokoyama, E., Putnis, C.V., Putnis, A., 2004. In situ observation of replacement reaction of KBr-KCl-water system by laser Machzehnder real-time phase-shift interferometry. Nippon Kessho Seicho Gakkaishi 31, 146.

Nitsch, K.H., 1973. Neue Erkenntnisse zur Stabilität von Lawsonit. Fortschr. Mineral 50, 34–35.

Nozaka, T., Fryer, P., 2011. Alteration of the oceanic lower crust at a slow-spreading axis: insight from vein-related zoned halos in olivine gabbro from Atlantis Massif, Mid-Atlantic Ridge. J. Petrol. 52, 643–664.

O'Neil, J.R., Clayton, R.N., Mayeda, T.K., 1969. Oxygen isotope fractionation in divalent metal carbonates. J. Chem. Phys. 51, 5547–5558.

O'Neil, J.R., Taylor Jr., H.P., 1967. The oxygen isotope and cation exchange chemistry of feldspars. Am. Mineral. 52, 1414–1437.

Oberti, R., Ottolini, L., Della Ventura, G., Parodi, G.C., 2001. On the symmetry and crystal chemistry of britholite: new structural and microanalytical data. Am. Mineral. 86, 1066–1075.

Ondruš, P., Veselovský, F., Hloušek, J., Skála, R., Vavrín, I., Frýda, J., Čejka, J., Gabašová, A., 1997. Secondary minerals of the Jáchymov (Joachimsthal) ore district. J. Czech. Geol. Soc. 42, 3–76.

Orozbaev, R., Hirajima, T., Bakirov, A., Takasu, A., Maki, K., Yoshida, K., Sakiev, K., Bakirov, A., Hirata, T., Tagiri, M., Togonbaeva, A., 2015. Trace element characteristics of clinozoisite pseudomorphs after lawsonite in talc-garnet-chloritoid schists from the Makbal UHP Complex, northern Kyrgyz Tian-Shan. Lithos 226, 98–115.

Pan, Y., Fleet, M.E., 2002. Compositions of the apatite-group minerals: substitution mechanisms and controlling factors. In: Kohn, M.J., Rakovan, J., Hughes, J.M. (Eds.), Phosphates: Geochemical, Geobiological and Materials Importance, vol. 48. Mineralogical Society of America, Washington, DC, pp. 13–49.

Pan, Y., Fleet, M.E., Macrae, N.D., 1993. Oriented monazite inclusions in apatite porphyroblasts from the Hemlo gold deposit, Ontario, Canada. Mineral. Mag. 57, 697–707.

Pankiwskyj, K.A., 1964. Geology of the Dixfield Quadrangle, Maine (Ph.D.). Harvard University.

Pearson, D.G., Davies, G.R., Nixon, P.H., 1995. Orogenic ultramafic rocks of UHP (diamond facies) origin. In: Coleman, R.G., Wang, X. (Eds.), Ultrahigh Pressure Metamorphism. Cambridge University Press, Cambridge, pp. 456–510.

Pearson, D.G., Davies, G.R., Nixon, P.H., Mattey, D.P., 1991. A carbon isotope study of diamond facies pyroxenites and associated rocks from the Beni Bousera peridotite, North Morocco. J. Petrol. (Sp. Lherzolites Issue) Special vol. 2, 175–189.

Pearson, D.G., Davies, G.R., Nixon, P.H., Milledge, H.J., 1989. Graphitized diamonds from a peridotite massif in Morocco and implications for anomalous diamond occurrences. Nature 335, 60–66.

Pedrosa, E.T., Putnis, C.V., Putnis, A., 2016. The pseudomorphic replacement of marble by apatite: the role of fluid composition. Chem. Geol. 425, 1–11.

Pekov, I.V., Siidra, O.I., Chukanov, N.V., Yapaskurt, V.O., Belakovskiy, D.I., Murashko, M.N., Sidorov, E.G., 2014. Kaliochalcite, $KCu_2(SO_4)_2[(OH)(H_2O)]$, a new tsumcorite-group mineral from the Tolbachik volcano, Kamchatka, Russia. Eur. J. Mineral. 26, 597–604.

Pelikan, A., 1902. Pseudomorphose von Magnetit und Rutil nach Ilmenit. Tschermaks Miner. Petrogr. Mitt. 21, 226–229.

Perdikouri, C., Kasioptas, A., Geisler, T., Schmidt, B.C., Putnis, A., 2011. Experimental study of the aragonite to calcite transition in aqueous solution. Geochim. Cosmochim. Acta 75, 6211–6224.

Perdikouri, C., Piazolo, S., Kasioptas, A., Schmidt, B.C., Putnis, A., 2013. Hydrothermal replacement of aragonite by calcite: interplay between replacement, fracturing and growth. Eur. J. Mineral. 25, 123–136.

Pewkliang, B., Pring, A., Brugger, J., 2008. The formation of precious opal: clues from the opalization of bone. Can. Mineral. 46, 139–149.

Pirajno, F., Grey, K., 2002. Chert in the palaeoproterozoic bartle member, killara formation, yerrida basin, Western Australia: a rift-related playa lake and thermal spring environment? Precambrian Res. 113, 169–192.

Pöllmann, H., Keck, E., 1993. Replacement and incrustation pseudomorphs of zeolites gismondite, chabazite, phillipsite and natrolite. Neues Jahrb. Mineral. Monatsh. 529–541.

Pollok, K., Putnis, C.V., Putnis, A., 2011. Mineral replacement reactions in solid solution-aqueous solution systems: volume changes, reactions paths and end-points using the example of model salt systems. Am. J. Sci. 311, 211–236.

Pöml, P., Menneken, M., Stephan, T., Niedermeier, D.R.D., Geisler, T., Putnis, A., 2007. Mechanism of hydrothermal alteration of natural self-irradiated and synthetic crystalline titanate-based pyrochlore. Geochim. Cosmochim. Acta 71, 3311–3322.

Porada, H., Behr, H.J., 1988. Setting and sedimentary facies of late Proterozoic alkali lake (playa) deposits in the southern Damara Belt of Namibia. Sediment. Geol. 58, 171–194.

Postma, D., 1985. Concentration of Mn and separation of Fe in sediments – kinetics and stoichiometry of the reaction between birnessite and dissolved Fe(II) at 10°C. Geochim. Cosmochim. Acta 49, 1023–1033.

Pring, A., Tenailleau, C., Etschmann, B., Brugger, J., Grguric, B., 2005. The transformation of pentlandite to violarite under mild hydrothermal conditions: a dissolution–reprecipitation reaction. In: Regolith, Ten Years of CRC LEME, pp. 252–255.

Puga, E., 1977. Sur l'existence dans le Complexe de la Sierra Nevada (Cordillère bétique, Espagne) d'éclogites et sur leur origine probable à partir d'une croîte océanique mésozoique. C.R. Acad. Sci. Sér. D 285, 1379–1382.

Putnis, A., 2002. Mineral replacement reactions: from macroscopic observations to microscopic mechanisms. Mineral. Mag. 66, 689—708.

Putnis, A., 2009a. Mineral replacement reactions. Rev. Mineral. Geochem. 70, 87—124.

Putnis, A., 2009b. Mineral replacement reactions. In: Oelkers, E.H., Schott, J. (Eds.), Thermodynamics and Kinetics of Water-Rock Interaction, vol. 70. Mineralogical Society of America, Chantilly, VA, pp. 87—124.

Putnis, A., Austrheim, H., 2010. Fluid-induced processes: metasomatism and metamorphism. Geofluids 10, 254—269.

Putnis, A., 2013. Mechanisms of metasomatism and metamorphism on the local mineral scale: the role of dissolution—reprecipitation during mineral reequilibration. In: Harlov, D.E., Austrheim, H. (Eds.), Metasomatism and the Chemical Transformation of Rock. Springer, Berlin, Heidelberg, pp. 141—170.

Putnis, A., Hinrichs, R., Putnis, C.V., Golla-Schindler, U., Collins, L.G., 2007a. Hematite in porous red-clouded feldspars: evidence of large-scale crustal fluid-rock interaction. Lithos 95, 10—18.

Putnis, A., John, T., 2010. Replacement processes in the Earth's crust. Elements 6, 159—164.

Putnis, A., Niedermeier, D.R.D., Putnis, C.V., 2006. From epitaxy to topotaxy: the migration of reaction interfaces through crystals. Geochim. Cosmochim. Acta 70, A509.

Putnis, A., Putnis, C.V., 2007. The mechanism of reequilibration of solids in the presence of a fluid phase. J. Solid State Chem. 180, 1783—1786.

Putnis, C.V., Austrheim, H., Putnis, A., 2007b. A mechanism of fluid transport through minerals. Geochim. Cosmochim. Acta 71, A814.

Putnis, C.V., Geisler, T., Schmid-Beurmann, P., Stephan, T., Giampaolo, C., 2007c. An experimental study of the replacement of leucite by analcime. Am. Mineral. 92, 19—26.

Putnis, C.V., Mezger, K., 2004. A mechanism of mineral replacement: isotope tracing in the model system KCl-KBr-H_2O. Geochim. Cosmochim. Acta 68, 2839—2848.

Putnis, C.V., Tsukamoto, K., Nishimura, Y., 2005. Direct observations of pseudomorphism: compositional and textural evolution at a fluid-solid interface. Am. Mineral. 90, 1909—1912.

Qian, G., Brugger, J., Skinner, W.M., Chen, G., Pring, A., 2010. Experimental study of the mechanism of the replacement of magnetite by pyrite up to 300°C. Geochim. Cosmochim. Acta 74, 5610—5630.

Qian, G., Pring, A., Brugger, J., Skinner, W.M., Chen, G., 2009. Replacement of magnetite by pyrite under hydrothermal conditions. J. Geochem. Explor. 101, 83.

Que, M., Allen, A.R., 1996. Sericitization of plagioclase in the Rosses Granite Complex, Co. Donegal, Ireland. Mineral. Mag. 60, 927—936.

Rendon-Angeles, J.C., Yanagisawa, K., Ishizawa, N., Oishi, S., 2000a. Conversion of calcium fluorapatite into calcium hydroxylapatite under alkaline hydrothermal conditions. J. Solid State Chem. 151, 65—72.

Rendon-Angeles, J.C., Yanagisawa, K., Ishizawa, N., Oishi, S., 2000b. Effect of metal ions of chlorapatite on the topotaxial replacement by hydroxylapatite under hydrothermal conditions. J. Solid State Chem. 154, 569—578.

Rendon-Angeles, J.C., Yanagisawa, K., Ishizawa, N., Oishi, S., 2000c. Topotaxial conversion of chlorapatite and hydroxylapatite to fluorapatite by hydrothermal ion exchange. Chem. Mater. 12, 2143—2150.

Richardson, S.W., 1968. Staurolite stability in a part of the system Fe-Al-Si-O-H. J. Petrol. 9, 467—488.

Robie, R.A., Bethke, P.M., Beardsley, K.K., 1967. Selected X-Ray Crystallographic Data, Molar Volumes, and Densities of Minerals and Related Substances. U.S. Government Printing Office.

Roering, C., Heckroodt, R.O., 1964. The alteration of beryl in the Dernburg pegmatite, Karibib, South West Africa. Ann. Geol. Surv. S. Afr. 3, 133—137.

Ruiz-Agudo, E., Putnis, C.V., Rodriguez-Navarro, C., Putnis, A., 2012. Mechanism of leached layer formation during chemical weathering of silicate minerals. Geology 40, 947—950.

Rybacki, E., Konrad, K., Renner, J., Wachmann, M., Stöckhert, B., Rummel, F., 2003. Experimental deformation of synthetic aragonite marble. J. Geophys. Res. Solid Earth 108, 8-1—8-15.

Šafanda, J., Szewczyk, J., Majorowicz, J., 2004. Geothermal evidence of very low glacial temperatures on the rim of the Fennoscandian ice sheet. Geophys. Res. Lett. 31, L07211.

Salvi, S., Williams-Jones, A.E., 1990. The role of hydrothermal processes in the granite-hosted zirconium, yttrium, REE deposit at Strange Lake, Quebec/Labrador: evidence from fluid inclusions. Geochim. Cosmochim. Acta 54, 2403—2418.

Salvi, S., Williams-Jones, A.E., 1996. The role of hydrothermal processes in concentrating HFSE in the Strange Lake peralkaline complex, northeastern Canada. Geochim. Cosmochim. Acta 60, 1917—1932.

Salvi, S., Williams-Jones, A.E., 2006. Alteration, HFSE mineralisation and hydrocarbon formation in peralkaline igneous systems: insights from the Strange Lake Pluton, Canada. Lithos 91, 19—34.

Salvi, S., Williams-Jones, A.E., Moine, B., 2000. Hydrothermal mobilization of high field strength elements in alkaline igneous systems: evidence from the Tamazeght Complex. Econ. Geol. 95, 559—576.

Sánchez-Pastor, N., Pina, C.M., Fernández-Díaz, L., 2007. A combined in situ AFM and SEM study of the interaction between celestite (001) surfaces and carbonate-bearing aqueous solutions. Surf. Sci. 601, 2973—2982.

Sawada, K., 1998. Mechanisms of crystal growth of ionic crystals in solution. Formation, transformation, and growth inhibition of calcium carbonates. In: Ohtaki, H. (Ed.), Crystallization Processes. Wiley, New York, pp. 39—68.

Scheerer, T., 1853. Ueber Pseudomorphosen, nebst Beiträgen zur Charakteristik einiger Arten Derselben. Poggendorff's Annalen der Physik und Chemie 89, 1—38.

Scheerer, T., 1854a. Der Paramorphismus und seine Bedeutung in der Chemie, Mineralogie und Geologie. Verlag von Friedrich Vieweg und Sohn, Braunschweig.

Scheerer, T., 1854b. Ueber Pseudomorphosen, nebst Beiträgen zur Charakteristik einiger Arten Derselben. Poggendorff's Analen der Physik und Chemie 92, 612—623.

Scheerer, T., 1854c. Ueber Pseudomorphosen, nebst Beiträgen zur Charakteristik einiger Arten Derselben. Poggendorff's Analen der Physik und Chemie 91, 378—400.

Scheerer, T., 1854d. Ueber Pseudomorphosen, nebst Beiträgen zur Charakteristik einiger Arten Derselben. Poggendorff's Analen der Physik und Chemie 93, 95—115.

Scheerer, T., 1856. Bemerkungen und Beobachtungen über Afterkrystalle. In: Liebig, J., Poggendorff, J.V., Wöhler, F. (Eds.), Handwörterbuche der reinen und angewandten Chemie, Volume Band 1. Druck und Verlag von Friedrich Vieweg und Sohn, Braunschweig, pp. 339–375.

Scheerer, T., 1953. Ueber Pseudomorphosen, nebst Beiträgen zur Charakteristik einiger Arten Derselben. Poggendorff's Analen der Physik und Chemie 90, 315–323.

Schuh, C.P., 2000. Mineralogy & Crystallography: A Biobibliography, 1469 to 1920. Curtis Schuh, Tucson, Arizona, 1356 p.

Schuiling, R.D., Vink, B.W., 1967. Stability relations of some titanium-minerals (sphene, perovskite, rutile, anatase). Geochim. Cosmochim. Acta 31, 2399–2411.

Shaikh, A.M., Shearman, D.J., 1986. On ikaite and the morphology of its pseudomorphs. In: Rodríguez-Clemente, R., Tardy, Y. (Eds.), Proc. Internat. Meeting on Geochemistry of the Earth Surface and Processes of Mineral Formation: Granada, Spain, Madrid: Consejo Superior de Investigaciones Cientificas. Centre national de la recherche scientifique, Paris, pp. 791–803.

Sinkankas, J., 1964. Mineralogy for Amateurs. D. Van Nostrand Company Inc., Princeton, NJ.

Slodkevich, V.V., 1980. Polycrystalline aggregates of octahedral graphite. Doklady 253, 194–196 (Translated from Doklady Akademii Nauk SSSR, 1980, Vol. 1253, No. 1983, 1697–1700).

Snow, E., Yund, R.A., 1987. The effect of ductile deformation on the kinetics and mechanisms of the aragonite–calcite transformation. J. Metamorph. Geol. 5, 141–153.

Storre, B., Nitsch, K.H., 1974. Zur Stabilität von Margarit in System CaO-Al$_2$O$_3$-SiO$_2$-H$_2$O. Contrib. Mineral. Petrol 43, 1–24.

Strunz, H., 1982. Pseudomorphosen-Der derzeitige Kenntnisstand. Versuch einer Klassifizierung. Aufschluss 33, 313–342.

Süsser, P., Schwertmann, U., 1983. Iron oxide mineralogy of ochreous deposits in drain pipes and ditches. Z. Kulturtech. Flurbereinig. 24, 386–395.

Trescases, J.J., 1973. Weathering and geochemical behaviour of the elements of ultramafic rocks in New Caledonia. Bur. Miner. Resour. Geol. Geophys. Dep. Miner. Energy Canberra Bull. 141, 149–161.

Velde, B., 1971. The stability and natural occurrence of margarite. Mineral. Mag 38, 317–323.

Villanova-de-Benavent, C., Proenza, J.A., Galí, S., García-Casco, A., Tauler, E., Lewis, J.F., Longo, F., 2014. Garnierites and garnierites: textures, mineralogy and geochemistry of garnierites in the Falcondo Ni-laterite deposit, Dominican Republic. Ore Geol. Rev. 58, 91–109.

von Knorring, O., 1963. VI(c) Report on Mineralogical Research, 7th Annual Report (1961–1962). Research Institute of African Geology, University of Leeds.

Warner, T.E., Rice, N.M., Taylor, N., 1996. Thermodynamic stability of pentlandite and violarite and new Eh-pH diagrams fro the iron-nickel-sulphur aqueous system. Hydrometallurgy 41, 107–118.

Wei, C., Powell, R., 2006. Calculated phase relations in the systemNCKFMASH (Na$_2$O-CaO-K$_2$O-FeO-MgO-Al$_2$O$_3$-SiO$_2$-H$_2$O) for high-pressure metapelites. J. Petrol. 47, 385–408.

Williams-Jones, A.E., Samson, I.M., Olivo, G.R., 2000. The genesis of hydrothermal fluorite-REE mineralization in the Gallinas Mountains, New Mexico. Econ. Geol. 95, 327–342.

Winkler, G., 1855. Die Pseudomorphosen des Mineralreichs. Joh. Palm's Hofbuchhandlung, Muenchen.

Wood, S.A., Samson, I.M., 2000. The hydrothermal geochemistry of tungsten in granitoid environments: I. Relative solubilities of ferberite and scheelite as a function of T, P, pH, and mNaCl. Econ. Geol. 95, 143–182.

Xia, F., Brugger, J., Chen, G., Ngothai, Y., O'Neill, B., Putnis, A., Pring, A., 2009a. Mechanism and kinetics of pseudomorphic mineral replacement reactions: a case study of the replacement of pentlandite by violarite. Geochim. Cosmochim. Acta 73, 1945–1969.

Xia, F., Ngothai, Y., Brugger, J., O'Neill, B., Chen, G., Pring, A., 2009b. Mechanism of pseudomorphic mineral replacement reactions revealed by a combined textural and kinetic study. J. Geochem. Explor. 101, 113.

Xiao, Y., Lasaga, A.C., 1994. Ab initio quantum mechanical studies of the kinetics and mechanisms of silicate dissolution H$^+$(H$_3$O$^+$) catalysis. Geochim. Cosmochim. Acta 58, 5379–5400.

Yanagisawa, K., Rendon-Angeles, J.C., Ishizawa, N., Oishi, S., 1999. Topotaxial replacement of chlorapatite by hydroxyapatite during hydrothermal ion exchange. Am. Mineral. 84, 1861–1869.

Zaraysky, G.P., 1989. Zonality and Conditions of Genesis of Metasomatic Rock. Nauka, Moscow, 341 p. (in Russian).

Zelenski, M.E., Zubkova, N.V., Pekov, I.V., Polekhovsky, Y.S., Pushcharovsky, D.Y., 2012. Cupromolybdite, Cu$_3$O(MoO$_4$)$_2$, a new fumarolic mineral from the Tolbachik volcano, Kamchatka Peninsula, Russia. Eur. J. Mineral. 24, 749–757.

Zhao, J., Brugger, J., Grundler, P.V., Xia, F., Chen, G., Pring, A., 2009. Mechanism and kinetics of a mineral transformation under hydrothermal conditions: calaverite to metallic gold. Am. Mineral. 94, 1541–1555.

2

Native Elements

SILVER PSEUDOMORPH AFTER DYSCRASITE

Silver pseudomorph after Dyscrasite,
Pribram, Bohemia-Czechoslovakia,
7.8 × 6 × 5.2 cm.

Dyscrasite,
Pribram, Bohemia, Czech Republic,
4.5 × 4.5 × 3.3 cm.

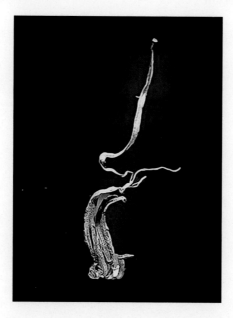

Silver,
Keeley-Frontier Mine, South Lorrain, Ontario, Canada,
5.75 cm tall.

Photo Atlas of Mineral Pseudomorphism
http://dx.doi.org/10.1016/B978-0-12-803674-7.00002-5

117

SILVER PSEUDOMORPH AFTER PYRARGYRITE

Silver pseudomorph after pyrargyrite,
Zacatecas, Mexico,
4.1 × 2.4 × 2.4 cm.

Pyrargyrite,
Andreasberg, Harz Mts., Germany,
2.2 × 1.5 × 1.2 cm.

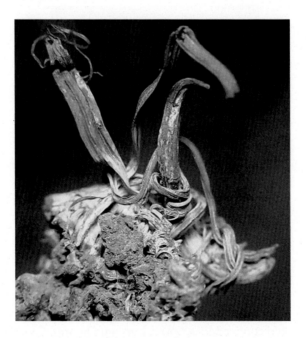

Silver on galena,
Homestake Mine, Creede, Colorado, USA,
2 × 1.5 × 1.5 cm.

COPPER PSEUDOMORPH AFTER ARAGONITE

Copper pseudomorph after aragonite cyclic twin,
Corocoro District, Pacajes Province, La Paz
Department, Bolivia,
2.5 × 2.5 × 1.8 cm.

"Pseudohexagonal" aragonite cyclic twin,
Molina de Aragon, Guadalajara, Castile-La
Mancha, Spain (type locality),
3.3 × 3.1 × 2.4 cm.

Copper,
Phoenix Mine, Keweenaw Co, Michigan, USA,
5.5 × 3 × 2.3 cm.

COPPER PSEUDOMORPH AFTER AZURITE

Copper pseudomorph after azurite,
Georgetown, Grant County, New Mexico, USA,
3.5 × 3.4 × 3 cm.

Copper, Centennial Mine,
Houghton Co., Michigan, USA,
4.0 × 2.5 × 2.4 cm.

Azurite,
Early 1960s: "Charlie Key pocket,"
Tsumeb, Namibia,
9.5 × 9 × 3 cm.

COPPER PSEUDOMORPH AFTER CUPRITE

Cuprite with malachite,
Musinoi Mine, Katanga, Zaire,
4.5 × 3.7 × 1.9 cm.

Copper pseudomorph after cuprite,
RubtsovskiyMine-Russia,
approximately 10 × 15 cm, bottom photo
field of view 5.5 cm and the large crystal
approximately 3 × 3 × 3 cm.

Copper twin,
Pewabic Lode Mines, Houghton County,
Michigan, USA,
4 × 4 × 2.8 cm.

3

Sulphides and Sulphosalts

ACANTHITE PSEUDOMORPH AFTER ARGENTITE

Acanthite pseudomorph after argentite,
argentite is only stable above 177°C and therefore all material is in the
stable form of acanthite at room temperature.
San Juan De Rayas Mine, Mexico,
2.5 cm tall × 2 cm × 2 cm.

Photo Atlas of Mineral Pseudomorphism
http://dx.doi.org/10.1016/B978-0-12-803674-7.00003-7

ARGENTITE–CHALCOPYRITE PSEUDOMORPH AFTER POLYBASITE

Acanthite–chalcopyrite pseudomorph after
polybasite,
Guanajuato, Mexico,
$5 \times 4.5 \times 2.25$ cm.

Polybasite,
Husky Mine, Elsa, Yukon Territory, Canada,
$2.2 \times 1.8 \times 0.4$ cm.

Acanthite,
Level 590, Sirena Mine, Guanajuato, Mexico,
$4.2 \times 2.5 \times 1$ cm.

Chalcopyrite,
Huanzala Mine, Huanuco Dept, Peru,
$3 \times 2.5 \times 1.7$ cm.

ACANTHITE PSEUDOMORPH AFTER PYRARGYRITE

Pyrargyrite,
Proano Mine, Fresnillo, Zacatecas, Mexico,
1.5 × 1.5 × 1.2 cm.

Acanthite pseudomorph after pyrargyrite,
Santa Catarina Mine, Guanajuato, Mexico,
7 × 7.2 × 5 cm.

Acanthite on calcite,
Uchucchacua Mine, Oyon Province, Lima
Department, Peru,
2.1 × 2.0 × 1.3 cm.

ACANTHITE PSEUDOMORPH AFTER GALENA

Acanthite pseudomorph after galena with
proustite and baryte,
Old Hope Of God Pit, Germany,
7 × 5 × 4.2 cm.

Galena with purple fluorite,
Elmwood Mine, Carthage, Tennessee, USA,
3 × 3 × 2 cm.

Acanthite on quartz,
Freiberg, Saxony, Germany,
5 × 4 × 3 cm.

CHALCOCITE PSEUDOMORPH AFTER COVELLITE

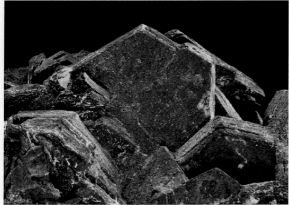

Chalcocite pseudomorph after covellite,
Leonard Mine, Montana, USA,
5 × 3.8 × 2.75 cm and the crystal is 2.3 cm
across.

Covellite,
Reynolds Tunnel, Summitville, Rio Grande
Co., Colorado, USA,
2.5 × 1.9 × 0.9 cm.

Chalcocite with calcite,
Bristol, Connecticut, USA,
6 × 4.5 × 3 cm.

CHALCOCITE PSEUDOMORPH AFTER GALENA

Chalcocite pseudomorph after galena,
Tsumeb, Namibia,
5.5 × 5 × 3.5 cm, crystal 2.6 cm.

Galena and pyrrhotite,
Dalnegorsk, Primorskiy Kray, Russia,
5 × 4 × 4.5 cm.

Chalcocite,
Flambeau Mine, Rusk County, Wisconsin, USA,
4 × 2.5 × 2.5 cm.

GALENA PSEUDOMORPH AFTER PYROMORPHITE

Galena pseudomorph after pyromorphite,
Poullaouen, France,
collected late-1700s–early 1800s!,
6 × 4 × 4 cm.

Pyromorphite,
Les Farges Mine, near Ussel, Correze, France,
5.5 × 5.5 × 4 cm.

Galena on calcite,
Sweetwater Mine, Ellington County, Missouri, USA,
6.5 × 5.5 × 3 cm.

SPHALERITE PSEUDOMORPH AFTER MAGNETITE

Martite (black iron rich sphalerite)
pseudomorph after magnetite,
Black Rock Mine, Beaver Co., Utah, USA,
6.8 × 6.0 × 5.0 cm.

Magnetite with orthoclase (adularia),
Binnental, Valais, Switzerland,
2.7× 2.2 × 1.5 cm.

Sphalerite twin on calcite,
Naica, Chihuahua, Mexico,
5.8 × 4.5 × 3 cm.

CHALCOPYRITE PSEUDOMORPH AFTER GALENA

Galena,
Joplin, Missouri, USA,
8.5 × 4.8 × 3.7 cm.

Chalcopyrite pseudomorph after galena,
Sweetwater Mine, Missouri, USA,
10 × 9 × 8 cm, crystals to 1.5 cm on edge.

Chalcopyrite with sphalerite,
Emperious Vein, Commodore Mine,
Creede, Mineral Co., Colorado, USA,
6.7 × 4.8 × 2.4 cm.

PYRITE PSEUDOMORPH AFTER PYRRHOTITE

Pyrite pseudomorph after pyrrhotite,
Huanzala Mine, Peru,
7.1 × 5.5 × 4.5 cm.

Pyrrhotite on quartz,
Dalnegorsk, Primorskiy Kray, Russia,
4.5 × 3.8 × 3.3 cm.

Pyrite,
Ibex Mine, Leadville, Colorado, USA,
5.7 × 4.9 × 4.5 cm.

SKUTTERUDITE PSEUDOMORPH AFTER SILVER

Silver,
North Kearsage Mine,
Houghton Co., Michigan, USA,
2.6 × 2.2 × 0.8 cm.

Skutterudite pseudomorph after silver,
Cobalt, Ontario, Canada,
7.9 × 5.1 × 3.4 cm.

Skutterudite,
Bou Azzer, Atlas Mountains, Morocco,
4.3 × 2.3 × 2 cm.

SAFFLORITE PSEUDOMORPH AFTER SILVER

Silver,
Wittichen, Baden, Black Forest, Germany,
2.7 × 2.2 × 0.2 cm.

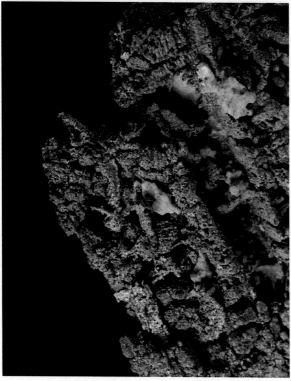

Safflorite pseudomorph after silver,
Langis Mine, Cobalt, Ontario, Canada,
8 × 7 × 4.5 cm.

Polished sample of safflorite, lollingite, and
rammelsbergite with quartz,
St. Andreasberg, Harz Mts., Lower Saxony,
Germany,
4.5 × 4.0 × 2.0 cm.

TENNANTITE PSEUDOMORPH AFTER AZURITE

Azurite,
early 1960s "Charlie Key pocket,"
Tsumeb, Namibia,
6 × 2 × 2 cm.

Tennantite pseudomorph after azurite,
Tsumeb Mine, Namibia,
6.1 × 6.0 × 4.9 cm.

Tennantite on quartz,
Tsumeb Mine, Tsumeb, Namibia,
11.5 × 11.0 × 8.0 cm.

TENNANTITE PSEUDOMORPH AFTER ENARGITE

Tennantite pseudomorph after enargite,
Julcani Mine, Peru,
8 × 6 × 4.5 cm.

Enargite,
Butte, Silver Bow County, Montana, USA,
4 × 3 × 1 cm.

Tennantite on quartz,
El Cobre Mine, Concepcion del Oro, Zacatecas, Mexico,
6.0 × 4.8 × 5.1 cm.

4

Oxides and Hydroxides

CORUNDUM PSEUDOMORPH AFTER SPINEL

Corundum pseudomorph after spinel,
Luc Yen, Myanmar (Burma),
5 × 4.7 × 4.5 mm.

Spinel,
Mogok, Mandalay, Myanmar (Burma),
1.3 × 1.0 × 0.5 cm.

Corundum, variety sapphire,
Ratnapura, Sri Lanka,
9.25 × 4 × 3.5 cm.

Photo Atlas of Mineral Pseudomorphism,
http://dx.doi.org/10.1016/B978-0-12-803674-7.00004-9

HEMATITE PSEUDOMORPH AFTER MAGNETITE

Hematite pseudomorph after magnetite,
Payun Volcano, Argentina,
11 × 4.75 × 4.25 cm.

Magnetite,
Cerro Huañaquino, Potosí Department,
Bolivia,
8.2 × 5.8 × 2.2 cm.

Hematite,
Novo Horizonte, Minas Gerais, Brazil,
3 × 2.2 × 2 cm.

HEMATITE PSEUDOMORPH AFTER MARCASITE

Hematite pseudomorph after marcasite,
White Desert, Egypt/Libya border,
Each approximately 3.5 cm across.

Marcasite,
Cape Blanc Nez, Calais, France,
11.6 × 10.5 × 8.7 cm.

Rutile epitaxial on hematite,
Novo Horizonte, Sao Paulo, Brazil,
5.2 × 4.0 × 1.2 cm.

RUTILE PSEUDOMORPH AFTER BROOKITE

Brookite on quartz,
Moses' Hill, Magnet Cove, Arkansas, USA,
1.7 × 1.6 × 1 cm.

Rutile pseudomorphs after brookite,
Magnet Cove, Arkansas, USA (Titanium
Corporation mine that was active between
1939 and 1944),
12 × 8 × 4.5 cm.

Rutile cyclic twin,
Magnet Cove, Hot Spring Co., Arkansas,
USA,
2.6 × 2.2 × 2 cm.

CASSITERITE PSEUDOMORPH AFTER ORTHOCLASE

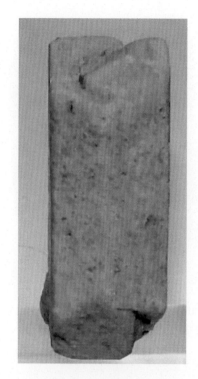

Orthoclase, Carlsbad twin,
Robinson Gulch Pegmatite, Jefferson Co.,
Colorado, USA,
4.5 × 2.7 × 1.8 cm.

Cassiterite pseudomorph after orthoclase
(Carlsbad twin) 1800's classic,
Wheal Coates, St. Agnes, Cornwall, England,
2.7 × 2.2 × 1.3 cm.

Cassiterite with fluorite,
Ehrenfriedersdorf, Erzgebirge, Saxony,
Germany,
5.1 × 3.4 × 3.4 cm.

GOETHITE PSEUDOMORPH AFTER CUPRITE

Goethite pseudomorph after cuprite,
Chessy, Lyon, Rhone-Alpes, France,
1.1 × 0.9 × 0.9 cm.

Cuprite,
Itauz Mine, Dzezhkazgan, Kazakhstan,
2 × 1.7 × 1.4 cm.

Goethite needles covered with tiny quartz crystals at the top on quartz matrix,
Atlas Mountains, Morocco,
4.4 × 3 × 2 cm.

GOETHITE PSEUDOMORPH AFTER GYPSUM

Goethite pseudomorph after gypsum,
Chihuahua, Mexico,
8.5 × 6 × 4.5 cm.

Gypsum,
nr. Gui Lin, Guangxi Province, China,
6.8 × 3.7 × 3 cm.

Goethite,
Crystal Creek, Lake George, Park Co., Colorado, USA,
8.5 × 5.4 × 4.3 cm.

GOETHITE PSEUDOMORPH AFTER PYRITE

Pyrite,
Huanzala Mine, Huanuco Dept., Peru,
20.5 × 13.0 × 6.3 cm.

Goethite pseudomorph after pyrite,
Pelican Point, Utah, USA,
26 cm high.

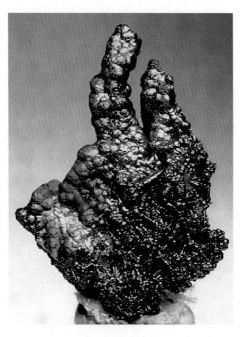

Goethite,
Garland County, Arkansas, USA,
2.6 × 1.8 × 0.8 cm.

GOETHITE PSEUDOMORPH AFTER SIDERITE

Goethite pseudomorph after siderite with
microcline (green variety amazonite),
Teller County, Colorado, USA,
17 × 15 × 12.5 cm.

Siderite,
Mt. St. Hilaire, Quebec, Canada,
4.0 × 2.7 × 4.3 cm.

Goethite,
Taouz, Morocco,
8.5 × 6.8 × 2.8 cm.

MANGANITE PSEUDOMORPH AFTER CALCITE

Calcite with tiny white blades of baryte,
Minerva #1 Mine, Harris Creek District,
Hardin County, Illinois, USA,
9.5 × 6 × 4 cm.

Manganite pseudomorph after calcite,
Harzburg near Ilfeld, Harz Mtns., Germany,
5 × 4.5 × 4 cm.

Manganite,
Ilfeld, Harz Mountains, Germany,
4 × 4.4 × 3.6 cm.

HAUSMANNITE PSEUDOMORPH AFTER MANGANITE

Hausmannite pseudomorph after manganite
with minor baryte,
Ilfeld, Nordhausen, Harz Mts, Thuringia,
Germany,
3.4 × 2.4 × 4.2 cm.

Manganite on quartz,
Ilfeld, Harz Mountains, Germany,
5.2 × 4.1 × 2.7 cm.

Hausmannite on andradite,
Wessels Mine, Kalahari Manganese Field, South Africa,
1.9 × 1.6 × 0.3 cm.

BIRNESSITE PSEUDOMORPH AFTER SERANDITE

Birnessite pseudomorph after serandite,
Mt. St. Hilaire, Canada,
7.2 × 6 × 4.5 cm.

Serandite,
Mont St Hilaire, Quebec, Canada,
2.4 × 0.7 × 1.2 cm.

Manganese biominerals (todorokite, romanechite, birnessite),
Chesapeake Bay, Calvert County, Maryland, USA,
9.1 × 4.3 × 2.3 cm.

CHAPTER

5

Halides

FLUORITE PSEUDOMORPH AFTER CALCITE

Calcite, Irai, Rio Grande do Sul, Brazil,
4.2 × 4 × 3.5 cm.

Fluorite pseudomorph after calcite,
Pili Mine, Mexico,
8 × 7 × 5 cm, crystal 3 cm tall.

Fluorite,
De An Mine, Jian Jiang, Jiangxi Province,
China,
5 × 4 × 2 cm.

Photo Atlas of Mineral Pseudomorphism
http://dx.doi.org/10.1016/B978-0-12-803674-7.00005-0

149

6

Carbonates

CALCITE PSEUDOMORPH AFTER ARAGONITE

Calcite pseudomorph after aragonite, with
sulfur,
Agrigento, Sicily, Italy,
8.5 × 7.3 × 5.2 cm.

Aragonite, variety tarnowitzite,
Tsumeb, Namibia,
8 × 4.5 × 1.5 cm.

Calcite on fluorite,
Minerva Mine, Cave-In-Rock District, Hardin Co., Illinois, USA,
8 × 7.5 × 6 cm.

Photo Atlas of Mineral Pseudomorphism,
http://dx.doi.org/10.1016/B978-0-12-803674-7.00006-2

151

CALCITE PSEUDOMORPH AFTER BARYTE

Calcite pseudomorph after baryte,
Santa Eulalia, Mexico,
13 × 9.5 × 8.5 cm.

Baryte,
Cerro Warihuyn, Miraflores Huamalias,
Huanuco, Peru,
6 × 5 × 3.8 cm.

Calcite with chalcopyrite-covered calcite phantom,
Kjorholt, Norway,
3.5 × 1.7 × 1.5 cm.

CALCITE PSEUDOMORPH AFTER FLUORITE

Fluorite,
De An Mine, Jian Jiang, Jiangxi Province,
China,
2.1 × 1.7 × 1.1 cm.

Calcite pseudomorph after fluorite with
quartz coating,
Dalnegorsk, Russia,
10.9 × 9 × 7.8 cm.

Calcite on mottramite,
Tsumeb Mine, Tsumeb, Namibia,
4.0 × 3.1 × 2.7 cm.

CALCITE PSEUDOMORPH AFTER IKAITE

Calcite,
Elmwood Mine, Carthage, Tennessee, USA,
12 × 9.5 × 9 cm.

Calcite pseudomorph after ikaite,
Olenitsa River, White Sea Coast, Karelia
Republic, Russia,
7.8 × 4.7 × 4.6 cm.

Calcite,
Joplin, Missouri, USA,
7.4 × 6 × 5.4 cm

Ikaite, $CaCO_3 \cdot 6H_2O$, dehydrates to calcite,
$CaCO_3$, above 8°C. The resulting
pseudomorph of calcite after ikaite is also
known as "glendonite."

CALCITE PSEUDOMORPH AFTER GLAUBERITE

Calcite pseudomorph after glauberite,
Camp Verde District, Yavapai Co., Arizona,
USA,
4.5 × 3.8 × 3.6 cm.

Glauberite,
Bertram Mine (Bertram siding sulfate
deposit; Bertram sodium sulfate deposits),
Bertram siding, Imperial Co., California,
USA,
4.1 × 2.9 × 2.3 cm.

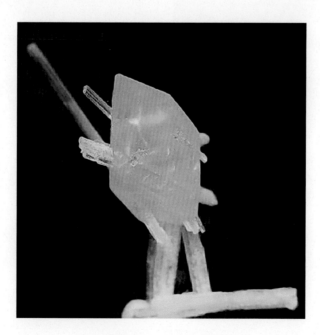

Calcite on scolecite,
Jalgaon, Maharashtra, India,
4 × 3 × 2 cm (including the points).

MAGNESITE PSEUDOMORPH AFTER CALCITE

Calcite,
Anderson Rock Products Quarry, Madison
County, Indiana, USA,
5.5 × 4 × 2.5 cm.

Magnesite pseudomorph after calcite,
Hohe Tauern Mts., Salzburg, Austria,
8.8 × 7.8 × 4.8 cm.

Magnesite,
Brumado, Bahia, Brazil,
6.9 × 6.4 × 5.7 cm.

SIDERITE PSEUDOMORPH AFTER APATITE

Siderite pseudomorph after apatite,
Panasquiera Mine, Portugal,
8.2 × 6 × 4.8 cm.

Fluorapatite with beryl, var. aquamarine, on muscovite,
Nagar, Pakistan,
8.6 × 6.7 × 4.8 cm.

Siderite,
Mont Saint-Hilaire, Rouville County, Quebec, Canada,
2.9 × 2.5 × 1.7 cm.

SIDERITE PSEUDOMORPH AFTER CALCITE

Siderite pseudomorph after calcite,
Turt Mine, Muramures, Romania,
9 × 7 × 7 cm.

Glass-clear calcite crystal on calcite rhomb
covered with quartz,
Nasik, India,
2 × 1.5 × 1.5 cm.

Siderite,
Estano Horco mine, Colavi District, Potosi Department, Bolivia,
4.0 × 3.8 × 2.2 cm.

SIDERITE PSEUDOMORPH AFTER SERANDITE

Siderite pseudomorph after serandite,
Mt. St. Hilaire, Quebec, Canada,
4.3 × 2.1 cm.

Serandite,
Mt. St. Hilaire, Quebec, Canada,
6 × 3× 2.5 cm.

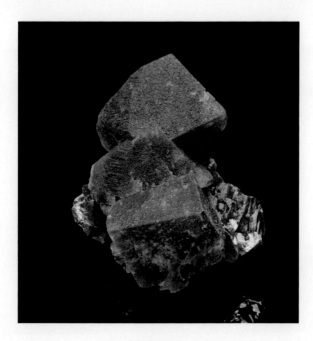

Siderite,
Tazna Mine, Cerro Tazna, Atocha-Quechisla District, Nor Chichas Province, Potosi
Department, Bolivia,
4 × 3 × 3 cm.

SIDERITE PSEUDOMORPH AFTER SPHALERITE

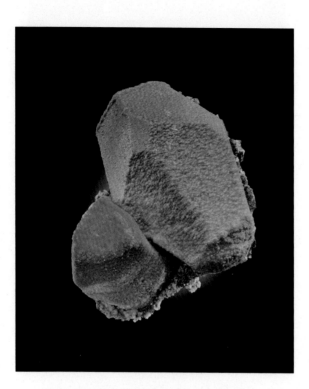

Siderite pseudomorph after sphalerite,
Aggenys Mine, South Africa,
6 × 6 × 5 cm.

Sphalerite with quartz,
Shuikoushan, Hunan Province, China,
9.1 × 5.5 × 4 cm.

Siderite on quartz,
Allevard, Isere, France,
9 × 7.5 × 3 cm.

RHODOCHROSITE PSEUDOMORPH AFTER CALCITE

Rhodochrosite pseudomorph after calcite,
Oppu Mine, Japan,
13.7 × 9 × 8.5 cm.

Cobaltian calcite with malachite,
Mashamba West, Kolwezi, Western area,
Shaba Cu belt, Shaba (Katanga), Congo
(Zaïre),
3.5 × 2.7 × 2.3 cm.

Rhodochrosite with quartz and fluorite,
Fourball Pocket, Fluorite Raise, Sweet
Home Mine, Alma, Colorado, USA,
5.5 × 3 × 2 cm.

RHODOCHROSITE PSEUDOMORPH AFTER FLUORITE

Water-clear fluorite (1.2 cm crystal) perched atop a matrix of quartz stalks, with some of them penetrating the fluorite, Dalnegorsk, Primorskiy Kray, Russia, 1.2 cm crystal.

Rhodochrosite pseudomorph after fluorite, Mt. St. Hilaire, Canada, 2.9 × 2.7 × 2.5 cm crystal.

Rhodochrosite, N'Chwaning Mine, Kalahari Manganese Field, Northern Cape Province, South Africa, 3.5 × 2.9 × 2.0 cm.

RHODOCHROSITE PSEUDOMORPH AFTER SERANDITE

Rhodochrosite pseudomorph after serandite,
Mt. St. Hilaire, Quebec, Canada,
6 × 3 × 2.5 cm.

Serandite,
Mt. St. Hilaire, Quebec, Canada,
4.5 × 3 × 2 cm.

Rhodochrosite with minor fluorite,
Sweet Home mine, Alma, Park County, Colorado, USA,
2.6 × 1.9 × 1.3 cm.

RHODOCHROSITE PSEUDOMORPH AFTER STURMANITE

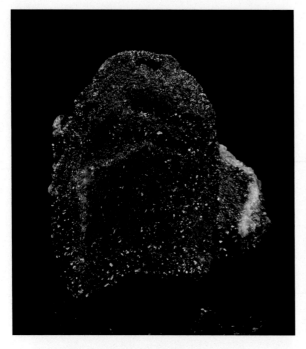

Rhodochrosite pseudomorph after
sturmanite,
NChwaning Mine, Kalahari Manganese
Fields, South Africa,
4 × 4 × 3 cm.

Sturmanite,
NChwaning Mine, Kalahari Manganese
Fields, South Africa,
1.5 cm tall crystal on matrix.

Rhodochrosite,
NChwaning II Mine, Kalahari Manganese Fields, South Africa,
2.7 × 1.5 × 1.25 cm.

SMITHSONITE PSEUDOMORPH AFTER ARAGONITE

Smithsonite pseudomorph after aragonite,
Tsumeb, Namibia,
12.5 × 9.5 × 9 cm.

Aragonite,
Bilin, Bohemia, Czech Republic,
3 × 1.7 × 0.9 cm.

Smithsonite,
Tsumeb, Namibia,
5.7 × 5.2 × 2.4 cm.

SMITHSONITE PSEUDOMORPH AFTER AZURITE

Smithsonite pseudomorph after azurite,
Tsumeb, Namibia,
5.5 × 4 × 3.5 cm.

Smithsonite (cobaltian),
El Refugio Mine, Sinaloa, Mexico,
5.7 × 5.3 × 2.5 cm.

Azurite on cerussite,
Tsumeb, Namibia,
6 × 6 × 3.5 cm.

SMITHSONITE PSEUDOMORPH AFTER CALCITE

Smithsonite pseudomorph after calcite,
San Antonio Mine, Chihuahua, Mexico,
6.7 × 6 × 5 cm.

Calcite twin on sphalerite,
Elmwood Mine, Carthage, Tenessee, USA,
10.5 × 7 × 4.5 cm.

Cuprian smithsonite,
Tsumeb, Namibia,
3.5 × 3 × 2.5 cm.

SMITHSONITE PSEUDOMORPH AFTER DOLOMITE

Smithsonite pseudomorph after dolomite,
Color due to greenockite inclusions,
Philadelphia Mine, Arkansas, USA,
8 × 6 × 4.5 cm.

Dolomite,
Binkley and Ober Quarry, Lancaster co.,
Pennsylvania, USA,
11.3 × 9.3 × 2.5 cm.

Manganoan smithsonite,
Tsumeb, Namibia,
8.5 × 5 × 5 cm.

SMITHSONITE PSEUDOMORPH AFTER DOLOMITE

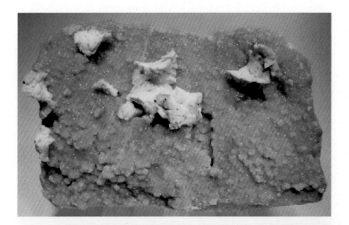

Dolomite,
Eugui, Navarra, Spain,
5.9 × 5.0 × 3.3 cm.

Smithsonite pseudomorph after dolomite on
quartz on silicified limestone,
Philadelphia Mine, Rush, Marion Co.,
Arkansas, USA,
8.7 × 5.3 × 3.2 cm.

Smithsonite on aurichalcite,
Kelly Mine, Magdalena, Socorro County,
New Mexico, USA,
4.1 × 3.6 × 1.8 cm.

SMITHSONITE PSEUDOMORPH AFTER SPHALERITE

Smithsonite pseudomorph after sphalerite
with dolomite,
Philadelphia Mine, Arkansas, USA,
13.6 × 9 × 8 cm.

Sphalerite on quartz,
Shuikoushan Lead-Zinc Mine, Chaling,
Hunan Province, China,
9 × 5 × 3 cm.

Smithsonite,
Tsumeb Mine, Namibia,
3.8 × 2.7 × 2.7 cm.

STRONTIANITE PSEUDOMORPH AFTER CELESTINE

Strontianite pseudomorph after celestine,
Lime City, Ohio, USA,
5.1 × 2.7 × 2.7 cm.

Celestine,
Majunga, Madagascar,
8.1 × 7.8 × 4.4 cm.

Strontianite on magnesite,
Oberdorf-on-the-Lamming, Styria, Tyrol, Austria,
5.6 × 3.5 × 3.5 cm.

CERUSSITE PSEUDOMORPH AFTER ANGLESITE

Cerussite pseudomorph after anglesite,
Tsumeb, Namibia,
11 × 8 × 4.5 cm.

Cerussite,
Toussit Mine, Ouijda, Atlas Mtns., Morocco,
2.2 × 1.7 × 1.4 cm.

Cerussite,
Tsumeb, Namibia,
12 × 11 × 4 cm.

CERUSSITE PSEUDOMORPH AFTER DESCLOIZITE

Cerussite-mimetite-hemimorphite
pseudomorph after descloizite,
Chah Mileh Mine, Iran,
7 × 5 × 4.5 cm.

Descloizite,
Berg Aukus, Grootfontein District, Namibia,
5.3 × 4.0 × 3.4 cm.

Mimetite,
Tsumeb, Namibia,
2.6 × 2.4 × 1.7 cm.

Hemimorphite,
San Antonio El Grande Mine, Chihuahua,
Mexico,
5.5 × 4 × 3 cm.

Cerussite with phantom,
Tsumeb, Namibia,
3 × 2.5 × 1.25 cm.

DOLOMITE PSEUDOMORPH AFTER ARAGONITE

Dolomite (cobaltian) pseudomorph after
aragonite,
Tsumeb, Namibia,
4.8 × 3.5 × 3.5 cm.

Aragonite,
Northern Lights Mine, Mineral County,
Nevada, USA,
4.2 × 4.0 × 3.7 cm.

Dolomite,
Eugui, Navarre, Spain,
9.6 × 7.5 × 4.5 cm.

DOLOMITE PSEUDOMORPH AFTER CALCITE

Dolomite pseudomorph after calcite,
Tsumeb Mine, Namibia,
9.5 × 8 × 8 cm.

Calcite,
Pau, Pyrénées-Atlantiques, Aquitaine, France,
4.0 × 3.7 × 3.0 cm.

Dolomite,
Binnnental, Switzerland,
2 × 1.8 × 1 cm.

AZURITE PSEUDOMORPH AFTER CUPRITE

Cuprite,
Dzhezkazgan, Kazakhstan,
1.5 × 1.3 × 1.1 cm.

Azurite
Tsumeb, Namibia,
11 × 10 × 7 cm.

Azurite pseudomorph after cuprite,
Chessy copper mines, Chessy-les-Mines, Le
Bois d'Oingt, Rhône, Rhône-Alpes, France,
1.4 × 1.3 × 0.8 cm.

AZURITE PSEUDOMORPH AFTER GYPSUM

Gypsum,
Santa Eulalia, Chihuahua, Mexico,
10 × 6.8 × 6.9 cm.

Azurite,
Tsumeb, Namibia,
4.8 × 2.6 × 2 cm.

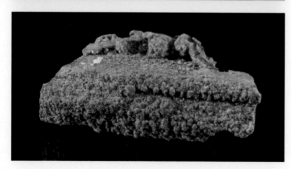

Azurite and malachite pseudomorph after
gypsum,
Apex Mine, Jarvis Peak, Beaver Dam Mts,
Washington Co.Utah, USA,
6.4 × 3.1 × 1.8 cm.

Malachite,
Bisbee, Cochise County, Arizona, USA,
7.8 × 6.5 × 3.3 cm.

HYDROCERUSSITE PSEUDOMORPH AFTER CERUSSITE

Cerussite,
Tsumeb Mine, Namibia,
3.6 × 3.2 × 1.6 cm.

Twinned hydrocerussite,
Tsumeb Mine, Tsumeb, Namibia,
3.6 × 3.5 × 2.2 cm.

Hydrocerussite pseudomorph after cerussite,
Tsumeb Mine, Namibia,
5.7 × 1.9 × 1.5 cm.

MALACHITE PSEUDOMORPH AFTER AZURITE

Azurite on anglesite,
Tsumeb, Namibia,
3.8 × 3 × 3 cm.

Malachite pseudomorph after azurite,
Campbell Mine, Bisbee, Arizona, USA,
9 × 8 × 6.5 cm.

Malachite with azurite nodule (polished),
Bisbee, Cochise County, Arizona,
USA,
5.2 × 4.9 × 2.0 cm.

MALACHITE PSEUDOMORPH AFTER BARYTE

Malachite pseudomorph after baryte,
Shangulowe Mine, Kambove District,
Katanga, Democratic Republic of Congo,
2.5 × 2.0 × 1.8 cm.

Baryte,
Huarihauyin, Miraflores, Huamalias,
Huanuco, Peru,
4 × 3 × 1.8 cm.

Malachite (primary) on quartz-coated chrysocolla,
Bagdad Mine, Bagdad, Arizona, USA,
2.8 × 2.5 × 0.9 cm.

ROSASITE PSEUDOMORPH AFTER MALACHITE

Rosasite after malachite after azurite with
cerussite,
Tsumeb, Namibia,
5 × 4 × 4 cm.

Azurite,
Czar Shaft, Bisbee, Cochise County,
Arizona, USA,
5.0 × 4.0 × 2.5 cm.

Rosasite on smithsonite,
O-22 area, Sherman mine, Leadville, Lake
County, Colorado, USA,
3.7 × 3.6 × 2.1 cm.

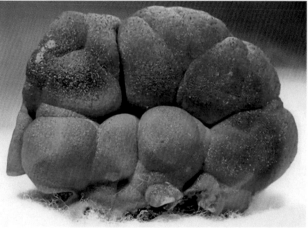

Malachite,
Kolwezi Mine, Kolwezi District, Katanga Copper
Crescent, Katanga (Shaba), Democratic Republic of
Congo (Zaïre),
4.9 × 3.7 × 3.1 cm.

LEADHILLITE PSEUDOMORPH AFTER CERUSSITE

Leadhillite on and pseudomorph after
cerussite,
Mammoth-St. Anthony Mine, Tiger, Pinal Co.,
Arizona, USA,
5.8 × 3.8 × 3.4 cm.

Cerussite,
Tsumeb, Namibia,
4.9 × 4.4 × 2.2 cm.

Leadhillite,
Tiger, Arizona, USA,
3.2 × 3.1 × 2.1 cm.

7

Borates

MEYERHOFFERITE PSEUDOMORPH AFTER INYOITE

Meyerhofferite pseudomorph after inyoite,
Monte Azul, Plano, Argentina,
17.7 × 9.2 × 8.1 cm.

Inyoite,
Monte Azul, Plano, Argentina,
6.3 × 6.0 × 5.5 cm.

Meyerhofferite,
Mount Blanco Mine, Death Valley, Inyo Co., California, USA (type locality),
4.0 × 2.7 × 2.2 cm.

BORAX PSEUDOMORPH AFTER HANKSITE

Borax (changing to tincalconite, starts
immediately after crystallization)
pseudomorph after hanksite,
Searles Lake, near Trona, San Bernardino Co.,
California, USA,
7.5 × 5.8 × 4.2 cm.

Hanksite,
Searles Lake, Trona, San Bernadino County,
California, USA (type locality),
8.4 × 4.7 × 4.5 cm.

TINCALCONITE PSEUDOMORPH AFTER BORAX

Tincalconite pseudomorph after borax,
US Borax Mine, California, USA,
96 mm.

Tincalconite (Artificial),
from mixer vats, Boron pit, Boron, Kramer District, Kern Co., California, USA,
7.0 × 4.2 × 3.8 cm.

HILGARDITE PSEUDOMORPH AFTER BORACITE

Hilgardite pseudomorph after boracite,
Boulby Potash Mine, Loftus, North Yorkshire,
England, UK,
5.5 × 3.5 × 2.8 cm.

Boracite,
Kalkberg, Lüneburg, Lower Saxony, Germany
(type locality),
1 × 1 × 0.8 cm.

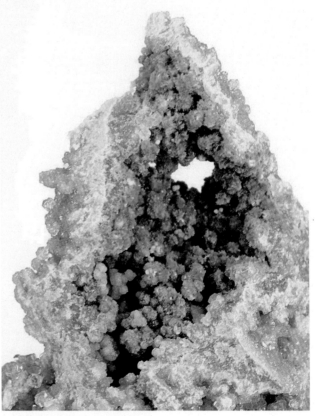

Hilgardite with boracite,
215 panel, Boulby Potash Mine, Loftus, Cleveland, North Yorkshire, England,
6.4 × 4.7 × 2.7 cm.

CHAPTER

8

Sulphates

THENARDITE PSEUDOMORPH AFTER MIRABILITE

Thenardite pseudomorph after mirabilite (as a
dehydration product),
Boron, Kramer District, Kern Co., California, USA,
6.4 × 5.0 × 4.0 cm.

Thenardite,
Soda Lake, Carizzo Plains, San Luis Obispo County,
California, USA,
9.6 × 6.4 × 5.8 cm.

ANGLESITE PSEUDOMORPH AFTER GALENA

Anglesite pseudomorph after galena,
Blanchard Mine, Bingham, New Mexico, USA,
2.5 × 2.0 × 1.7 cm.

Galena with fluorite,
Elmwood Mine, Carthage, Tennessee, USA,
3 × 3 × 2 cm.

Anglesite,
Tsumeb, Namibia,
4.4 × 4 × 2.3 cm.

GYPSUM PSEUDOMORPH AFTER CALCITE

Calcite twin,
Elmwood Mine, Carthage, Tennessee, USA,
9 × 6 × 3.7 cm.

Gypsum pseudomorph after calcite with
melanterite on sphalerite,
Mid-Continent Mine, Picher, Oklahoma (old
"Tri-State District"), USA,
7 × 5 × 4.5 cm.

Gypsum,
Roden, Zaragoza, Spain,
5.5 × 4.9 × 4.5 cm.

GYPSUM PSEUDOMORPH AFTER GLAUBERITE

Gypsum pseudomorph after glauberite,
Camp Verde, Camp Verde District, Yavapai
Co., Arizona, USA,
4.9 × 3.1 × 2.9 cm.

Glauberite,
Camp Verde, Yavapai County, Arizona,
5.2 × 2.8 × 2.7 cm.

Calcite,
Irai, Rio Grande Do Sul, Brazil,
4 × 4 × 3.5 cm.

BROCHANTITE PSEUDOMORPH AFTER BORNITE

Bornite on quartz,
Jezkazgan, Kazakhstan,
3.8 × 2.4 × 1.5 cm.

Brochantite pseudomorph after bornite,
Jezkazgan, Karaganda Oblast, Kazakhstan,
9 × 7 × 5.8 cm.

Brochantite with blue linarite on quartz,
Blanchard Mine, Bingham, Socorro Co., New
Mexico, USA,
5.5 × 4.2 × 2.8 cm.

LINARITE PSEUDOMORPH AFTER GALENA

Linarite and anglesite pseudomorph after galena,
Blanchard Mine, Socorro, New Mexico, USA,
6 × 4 × 3.7 cm.

Galena,
Joplin, Missouri, USA,
8.5 × 4.8 × 3.7 cm.

Anglesite,
Touissit Mine, Ouijda, Morocco,
7 × 3.5 × 2.2 cm.

Linarite on a matrix of tabular barite and
some fluorite,
Sunshine #1 Adit, Blanchard Mine,
Bingham, Socorro Co., New Mexico, USA,
7.5 × 6.2 × 6.0 cm.

POSNJAKITE PSEUDOMORPH AFTER CALCITE

Posnjakite and langite pseudomorph after
calcite,
Canaveille, Pyrenees, France,
8 × 6 × 5 cm.

Calcite on quartz,
Elmwood Mine, Carthage, Tennessee,
3.3 × 2.8 × 1.5 cm.

Posnjakite,
Drakewells Mine, Gunnislake, Cornwall,
England,
11.0 × 7.5 × 4.2 cm.

Langite,
Podlipa Mine, Lubietova, Neusohl District,
Slovak Republic,
5.7 × 5.3 × 1.7 cm.

STURMANITE PSEUDOMORPH AFTER BARYTE

Sturmanite pseudomorph after baryte,
Black Rock Mine, near Kuruman, South Africa,
2.1 × 1.3 × 1 cm.

Sturmanite,
N'Chwaning II Mine, Kalahari Manganese
Fields, South Africa,
3.1 × 2.8 × 2.1 cm.

Baryte,
Stoneham, Weld County, Colorado, USA,
4.5 × 3.2 × 2.7 cm.

STURMANITE PSEUDOMORPH AFTER GYPSUM

Sturmanite pseudomorph after gypsum,
Black Rock Mine, near Kuruman, South Africa,
2.5 × 1 × 0.5 cm.

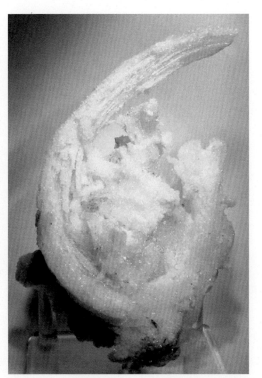

Ramshorn gypsum,
San Antonio Mine, Santa Eulalia,
Chihuahua, Mexico,
12.7 × 9.4 × 7.7 cm.

Sturmanite,
N'Chwaning Mine, Kalahari Manganese Fields, South Africa,
5.4 × 1.8 × 1.3 cm.

9

Phosphates, Arsenates, and Vanadates

NEWBERYITE PSEUDOMORPH AFTER STRUVITE

Newberyite pseudomorph after struvite,
Paoha Island, Mono Lake, Mono Co.,
California, USA,
3.1 × 2.7 × 0.7 cm.

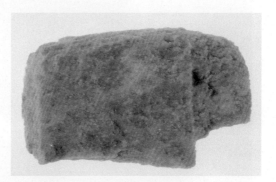

Struvite,
Skipton Lava Caves, Victoria, Australia,
3 × 1.8 × 1 cm.

Newberyite,
Skipton Lava Caves, Victoria, Australia (type locality),
3.2 × 2.5 × 1.9 cm.

Photo Atlas of Mineral Pseudomorphism
http://dx.doi.org/10.1016/B978-0-12-803674-7.00009-8

TSUMCORITE PSEUDOMORPH AFTER MIMETITE

Tsumcorite pseudomorph after mimetite,
Tsumeb, Namibia,
1st Oxidation Zone c.early 1900s,
4 × 4 × 3 cm.

Mimetite,
1971 Gem Pocket, Tsumeb, Namibia,
6.25 × 5 × 3 cm.

Tsumcorite,
Tsumeb, Namibia,
1.8 × 1.7 × 1.4 cm.

BAYLDONITE PSEUDOMORPH AFTER MIMETITE

Bayldonite pseudomorph after mimetite,
Tsumeb, Namibia,
5 × 4 × 3.5 cm.

Mimetite,
Cobar Mine, Cobar, New South Wales,
Australia,
2.7 × 2.5 × 1.8 cm.

Bayldonite with azurite,
Tsumeb, Namibia,
4.1 × 3.4 × 2.6 cm.

MOTTRAMITE PSEUDOMORPH AFTER CERUSSITE

Mottramite pseudomorph after cerussite,
Tsumeb, Namibia,
6.5 × 5 × 4 cm.

Cerussite,
Tsumeb, Namibia,
6.8 × 4.5 × 3.2 cm.

Mottramite,
Tsumeb, Namibia,
16 × 11 × 10 cm.

MOTTRAMITE PSEUDOMORPH AFTER COPPER

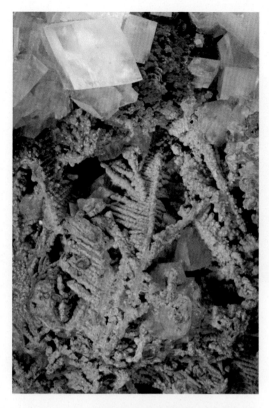

Mottramite pseudomorph after copper
with calcite,
Tsumeb, Namibia,
Close-up of a larger specimen. Field of
view approximately 2 cm.

Copper,
Bisbee, Cochise County, Arizona, USA,
3.0 × 2.8 × 2.6 cm.

Mottramite,
Tsumeb, Namibia,
6.7 × 4.8 × 3.2 cm.

MOTTRAMITE PSEUDOMORPH AFTER CALCITE

Mottramite pseudomorph after calcite,
Tsumeb Mine, Namibia,
3.0 × 3.0 × 1.3 cm.

Calcite with hematite inclusions,
Tsumeb, Namibia,
10 × 7 × 6 cm.

Mottramite,
Tsumeb Mine, Tsumeb, Namibia,
5.3 × 4.3 × 3.3 cm.

MOTTRAMITE PSEUDOMORPH AFTER WULFENITE

Mottramite pseudomorph after wulfenite on
plumbian calcite,
Tsumeb, Namibia,
5 × 4.5 × 3 cm.

Wulfenite,
Red Gem Pocket, Red Cloud Mine, La Paz
County, Arizona, USA,
12 × 9 × 5 cm.

Mottramite on calcite,
Tsumeb, Namibia,
25 × 25 × 16 mm.

PYROMORPHITE PSEUDOMORPH AFTER CERUSSITE

Cerussite,
Tsumeb, Namibia,
5.4 × 3.4 × 2.4 cm.

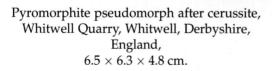

Pyromorphite pseudomorph after cerussite,
Whitwell Quarry, Whitwell, Derbyshire,
England,
6.5 × 6.3 × 4.8 cm.

Pyromorphite,
Daoping Lead-Zinc Mine, Gongcheng County,
Guangxi Province, China,
7.0 × 4.3 × 3.7 cm.

MIMETITE PSEUDOMORPH AFTER CERUSSITE

Cerussite twin,
Tsumeb Mine, Namibia,
$4.1 \times 3.0 \times 3.0$ cm.

Mimetite pseudomorph after cerussite,
Tsumeb, Namibia,
$6.8 \times 4.5 \times 4$ cm.

Mimetite,
1971 Gem Pocket, Tsumeb, Namibia,
$2.4 \times 1.9 \times 1.2$ cm.

MIMETITE PSEUDOMORPH AFTER ANGLESITE

Anglesite,
Touissit Mine, Ouijda, Morocco,
2.4 × 2.2 × 1.1 cm.

Mimetite pseudomorph after anglesite,
Tsumeb, Namibia,
5 × 3.25 × 2.5 cm.

Mimetite,
Tsumeb, Namibia,
2.1 × 1.6 × 1.4 cm.

MIMETITE PSEUDOMORPH AFTER VANADINITE

Mimetite pseudomorph after vanadinite
with dark olive green mottramite,
Touissit, Oujda, Morocco,
4.4 × 3 × 2.9 cm.

Vanadinite,
Touissit, Oujda, Morocco,
7 × 4.2 × 2.5 cm.

Mimetite,
San Pedro Corralitos, Chihuahua, Mexico,
2.3 × 2 × 0.9 cm.

MIMETITE PSEUDOMORPH AFTER WULFENITE

Mimetite pseudomorph after wulfenite,
San Francisco Mine, Sonora, Mexico,
4.7 × 4.3 × 4.2 cm.

Wulfenite,
Touissit Mine, Atlas Mountains, Morocco,
16.1 × 12.1 × 4.7 cm.

Mimetite,
Gem Pocket 1971, Tsumeb, Namibia,
2.6 × 1.7 × 1.2 cm.

VANADINITE PSEUDOMORPH AFTER WULFENITE

Vanadinite pseudomorph after wulfenite,
Red Cloud Mine, Arizona, USA,
1 cm crystal.

Wulfenite,
Rowley Mine, Arizona, USA,
$2.2 \times 2 \times 1$ cm.

Vanadinite
Mibladen, Morocco,
about $3 \times 3 \times 1.5$ cm.

PLUMBOGUMMITE PSEUDOMORPH AFTER PYROMORPHITE

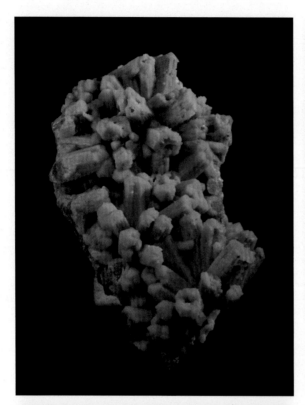

Pyromorphite,
Bunker Hill Mine, Kellogg, Shoshone Co.,
Idaho, USA,
3.5 × 2.25 × 2 cm.

Plumbogummite pseudomorphite after
pyromorphite,
Yangshuo Mine, China,
6.2 × 4.25 × 4.25 cm.

Plumbogummite,
Yangshuo Pb-Zz Mine, Yangshuo, Guangxi,
Province, China,
2.8 × 1.5 × 1.5 cm.

TURQUOISE PSEUDOMORPH AFTER APATITE

Turquoise pseudomorph after apatite,
Nacozari, Mexico,
4.5 × 2.5 × 2.5 cm.

Fluorapatite,
Cerro De Mercado, Durango, Mexico,
3.4 × 1.6 × 1.6 cm.

Crystallized turquoise on a matrix of quartz
and muscovite,
Lynch Station, Campbell Co., Virginia,
USA,
3.1 × 3.1 × 1.8 cm.

ARSENTSUMEBITE PSEUDOMORPH AFTER MIMETITE

Arsentsumebite pseudomorph after
mimetite,
Tsumeb, Namibia,
7.1 × 6 × 5 cm.

Mimetite with green duftite,
Tsumeb Mine, Namibia,
6.6 × 6.1 × 3 cm.

Arsentsumebite with azurite,
Tsumeb, Namibia,
3.7 × 2.8 × 2.4 cm.

10

Antimonates

STIBICONITE PSEUDOMORPH AFTER STIBNITE

Stibiconite pseudomorph after stibnite,
Tanger Mine, Mexico,
26 cm.

STIBICONITE PSEUDOMORPH AFTER STIBNITE

Stibnite,
Wuling Antimony Mine, Jiangxi Province,
China, 11.9 × 8.7 × 3.6 cm.

11

Molybdates and Tungstates

FERBERITE PSEUDOMORPH AFTER SCHEELITE

Ferberite pseudomorph after scheelite,
Long Hill, Haddam, Middlesex Co.,
Connecticut, USA,
2.0 × 2.0 × 1.5 cm.

Scheelite on muscovite,
Mt. Xuebaoding, Ping Wu, Sichuan Province,
China,
4.5 × 3.5 × 3 cm.

Ferberite,
Tasna mine, Rosario section of Cerro Tasna, Nor Chichas Province, Potosi Department,
Bolivia,
11.5 × 6.0 × 4.5 cm.

Photo Atlas of Mineral Pseudomorphism
http://dx.doi.org/10.1016/B978-0-12-803674-7.00011-6

WULFENITE PSEUDOMORPH AFTER ANGLESITE

Wulfenite pseudomorph after anglesite,
Tsumeb, Namibia,
3.3 × 2.25 × 2.25 cm.

Anglesite,
Mibladen, Morocco,
8 × 5.5 × 5 cm.

Wulfenite with mimetite,
San Francisco Mine, Magdalena, Sonora, Mexico,
4 × 3.5 × 3.5 cm.

WULFENITE PSEUDOMORPH AFTER SCHEELITE

Wulfenite pseudomorph after scheelite,
Trumbull, Connecticut, USA,
2.5 × 2.5 × 2.5 cm.

Scheelite,
Tankergin, Siberia, Russia,
Approx. 4 × 4 × 4 cm.

Wulfenite,
Touissit Mine, Ouijda, Morocco,
3.2 × 2.4 × 2.3 cm.

CUPROTUNGSTITE PSEUDOMORPH AFTER SCHEELITE

Cuprotungstite pseudomorph after scheelite,
Gold Hill District (Clifton District), Deep Creek
Mts, Tooele Co., Utah, USA,
2.0 × 1.8 × 1.6 cm.

Scheelite on quartz,
Yaogangxian Mine, Chenzhou, Hunan
Province, China,
sample size: 9.3 × 5.2 × 1.4 cm.

Cuprotungstite,
Gamsberg, Khomas Region, Namibia,
5.2 × 3.4 × 3.0 cm.

12

Nesosilicates

ANDRADITE PSEUDOMORPH AFTER RHODOCHROSITE

Andradite pseudomorph after rhodochrosite,
N'Chwaning, Kuruman, Kalahari manganese
field, Northern Cape Province, South Africa,
5.5 × 5 × 1.8 cm.

Rhodochrosite,
N'Chwaning Mine, Kalahari, South Africa,
3 × 2 × 1.75 cm.

Andradite, variety demantoid,
origin: Tubussis farm, Erongo, Namibia,
7 × 5 × 2 cm.

EUCLASE PSEUDOMORPH AFTER BERYL

Euclase pseudomorph after beryl,
Lost Hope Mine, Miami, Karoi District,
Mashonaland West, Zimbabwe,
10 × 10 × 9 cm.

Beryl, variety aquamarine, on quartz and
feldspar, with microlite inclusions (repaired
in the middle),
Bongala Springs Mine, Shigar Valley, Gilgit,
Pakistan,
32.5 × 22.9 × 12.8 cm.

Euclase,
Lost Hope Mine, Miami, Karoi District,
Mashonaland West, Zimbabwe,
2.6 × 1.9 × 1.8 cm.

TOPAZ PSEUDOMORPH AFTER ORTHOCLASE

Orthoclase with Carlsbad twinning,
Karlovy Vary (Carlsbad), Bohemia, Czech
Republic,
3.5 × 3.5 × 2 cm.

Topaz pseudomorph after orthoclase with
Carlsbad twinning,
Schneckenstein, Vogtland, Saxony, Germany,
3.0 × 2.1 × 1.2 cm.

Topaz,
Dassu, Gilgit, Pakistan,
6.8 × 5 × 2.5 cm.

FLUORELLESTADITE PSEUDOMORPH AFTER HEMIMORPHITE

Fluorellestadite pseudomorph after
hemimorphite,
Mapimi, Mexico,
Field of view 2.5 cm.

Hemimorphite,
Mapimi, Durango, Mexico,
10.6 × 6.9 × 6.2 cm.

Fluorellestadite in blue calcite,
Commercial Quarry, Sky Blue Hill, Crestmore
quarries, Crestmore, Riverside Co.,
California, USA,
5.1 × 4 × 2.7 cm.

DATOLITE PSEUDOMORPH AFTER ANHYDRITE

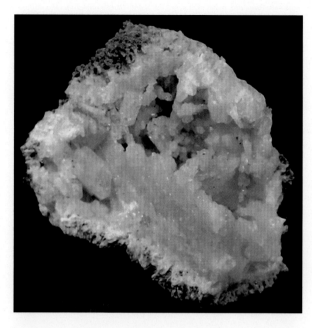

Datolite pseudomorph after anhydrite, dozens of light green datolite crystals forming a flat "casts" where a blade of anhydrite once existed, but has since dissolved away,
Upper New Street Quarry, Paterson, Passaic Count, New Jersey, USA,
9.0 × 7.3 × 5.7 cm.

Anhydrite,
Naica, Chihuahua, Mexico,
13.8 × 6.8 × 5.8 cm.

Datolite,
Verchniy Mine, Dalnegorsk, Russia,
3.5 × 2.7 × 2.2 cm.

13

Sorosilicates

HEMIMORPHITE PSEUDOMORPH AFTER CALCITE

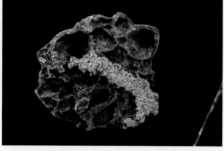

Hemimorphite pseudomorph
after calcite,
Joplin, Missouri, USA,
15.7 × 13 × 9 cm.

Calcite,
Pallaflat Mine, Cumbria, England,
11 × 9 × 4.5 cm.

Hemimorphite,
San Antonio El Grande Mine, Chihuahua,
Mexico,
5.5 × 4 × 3 cm.

Photo Atlas of Mineral Pseudomorphism
http://dx.doi.org/10.1016/B978-0-12-803674-7.00013-X

225

EPIDOTE PSEUDOMORPH AFTER MAGNETITE

Epidote pseudomorph after magnetite,
Raskoh Mts., Kharan, Balochistan
(Baluchistan), Pakistan,
9.1 × 6 × 5 cm.

Magnetite,
Cerro Huanaquino, Potosi Department,
Bolivia,
6.6 × 5.5 × 2.4 cm.

Epidote,
Prince of Wales Island, Alaska, USA,
3.2 × 2.2 × 1.8 cm.

14

Cyclosilicates

SCHORL PSEUDOMORPH AFTER BERYL

Schorl pseudomorph after beryl,
Erongo, Namibia,
8.6 × 6 × 4.5 cm.

Schorl on smoky quartz,
Erongo Mountains, Namibia,
7.5 × 6.2 × 5.4 cm.

Beryl, variety emerald,
Coscuez Mine, Boyaca Dept., Columbia, USA,
2.5 × 1.3 × 1 cm.

CHAPTER

15

Inosilicates

HEDENBERGITE PSEUDOMORPH AFTER ILVAITE

Hedenbergite pseudomorph after ilvaite,
Dalnegorsk, Russia,
7.3 × 4.8 × 4.3 cm.

Ilvaite,
Bor Mine, Primorjski Kraj Dalnegorsk,
Russia,
4.0 × 3.2 × 2.7 cm.

Hedenbergite,
Dalnegorsk, Russia,
9 × 7.7 × 4.6 cm.

Photo Atlas of Mineral Pseudomorphism
http://dx.doi.org/10.1016/B978-0-12-803674-7.00015-3

229

SHATTUCKITE PSEUDOMORPH AFTER CALCITE

Calcite,
Durango, Mexico,
24 × 14 × 11 cm.

Shattuckite pseudomorph after calcite with
dioptase,
Tantara Mine, Democratic Republic of
Congo,
5.9 × 4.4 × 4.25 cm.

Shattuckite,
Mesopotamia, Khorixas, Damaraland
District, Namibia,
2.9 × 2.6 × 2.1 cm.

SHATTUCKITE PSEUDOMORPH AFTER CUPRITE

Cuprite,
Tsumeb Mine, Tsumeb, Namibia,
2.3 × 2.1 × 1.2 cm.

Shattuckite pseudomorph after cuprite,
M'sesa Deposit, Kambove, Katanga,
Congo,
1.7 × 1.4 × 1.3 cm.

Dioptase on shattuckite in chrysocolla vug,
Otjikotu, Kaokoveld, Kunene, Namibia,
4.5 × 4.0 × 2.5 cm.

16

Phyllosilicates

ANTIGORITE PSEUDOMORPH AFTER CHONDRODITE

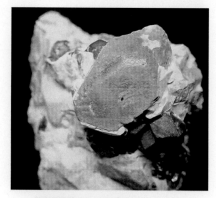

Antigorite (serpentine) pseudomorph after chondrodite with magnetite on kaolinite, Tilly Foster Mine, Brewster, Putnam County, New York, USA, 5 × 5 × 3 cm.

Chondrodite with magnetite, Tilly Foster Mine, Brewster, Putnam Co., New York, USA, 4.5 × 2.7 × 2.5 cm.

Antigorite, Tilly Foster Iron Mine, Putnam Co., New York, USA, 6.8 × 4.4 × 2.7 cm.

TALC PSEUDOMORPH AFTER QUARTZ

Talc pseudomorph after quartz,
Gopfersgrun, Germany,
9.8 × 8 × 6 cm.

Quartz,
Isere, France
6.9 × 5.1 × 2.4 cm.

Talc,
Gerendorf, Oberwald, Canton Wallis,
Switzerland,
6.7 × 4.8 × 2.8 cm.

TALC PSEUDOMORPH AFTER SPODUMENE

Talc pseudomorph after spodumene,
Greenwood, Oxford County, Maine, USA,
$17.0 \times 8.0 \times 6.0$ cm.

Spodumene, variety kunzite,
Nuristan, Laghman Province, Afghanistan,
$8.5 \times 4.6 \times 2.1$ cm.

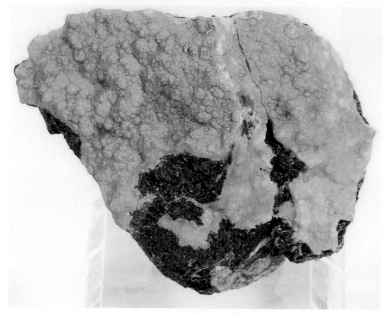

Antigorite, nickel-containing variety
previously known as genthite, Wood's
Chrome Mine (Wood's Mine), Texas, Little
Britain Township, Lancaster Co., Pennsylvania,
USA, $4.6 \times 3.9 \times 1.9$ cm.

MUSCOVITE PSEUDOMORPH AFTER GARNET

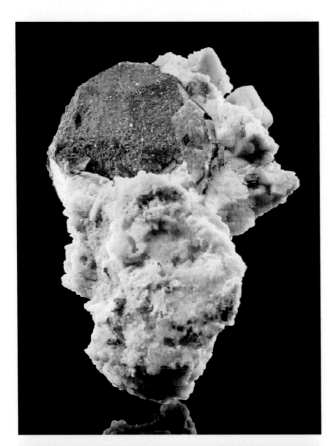

Grossular garnet, variety tsavorite,
Outokumpu, Finland,
1.75 × 1.5 × 1.25 cm.

Muscovite pseudomorph after garnet,
Urucum Mine, Minas Gerais, Brazil,
11.4 × 8 × 7.5 cm crystal is 7 cm across.

Muscovite,
Minas Gerais, Brazil,
5.9 × 5.1 × 3.6 cm.

POLYLITHIONITE PSEUDOMORPH AFTER TOURMALINE

Polylithionite (discredited name lepidolite)
pseudomorph after tourmaline,
Minas Gerais, Brazil,
6.5 cm tall

Tourmaline,
Minas Gerais, Brazil,
4.5 × 2.5 × 2.5 cm.

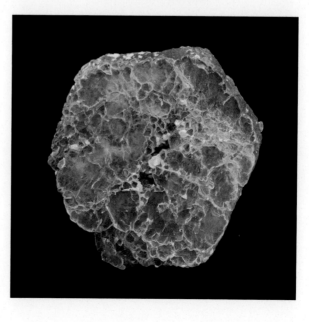

Polylithionite,
Minas Gerais, Brazil,
3 × 3 × 1 cm.

COOKEITE PSEUDOMORPH AFTER TOURMALINE

Cookeite pseudomorph after tourmaline,
Golconda Mine, Minas Gerais, Brazil,
15.4 × 7.5 × 7.5 cm.

Tourmaline with lepidolite,
origin: Governador Valaderes, Minas Gerais,
Brazil,
5.4 × 3.4 × 3.1 cm.

Green spherules of cookite with quartz,
Stand-on-Your-Head No. 1 mine, Bland,
Saline Co., Arkansas, USA,
8.5 × 6.7 × 4.3 cm.

CLINOCHLORE (CHLORITE) PSEUDOMORPH AFTER ALMANDINE

Clinochlore (chlorite) pseudomorph after
almandine,
Spurr Mine, Michigamme, Baraga Co., Michigan,
9.6 × 6.0 × 4.8 cm.

Almandine (garnet),
Minas Gerais, Brazil,
4.9 × 3.3 × 2.8 cm.

Clinochlore with smoky quartz,
Gletscheralp, Valais, Switzerland,
16.5 × 10.1 × 3.8 cm.

PREHNITE PSEUDOMORPH AFTER CALCITE

Calcite,
Durango, Mexico,
23 × 10 × 8 cm.

Prehnite pseudomorph after calcite,
Calvinia, near Capetown, Cape Province,
South Africa,
9 × 7 × 0.3 cm.

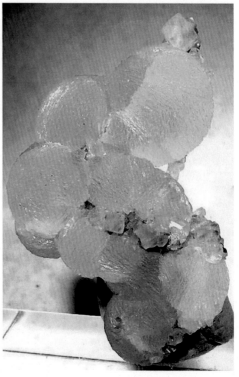

Prehnite with apophyllite,
Prospect Park Quarry, Passaic Co., New
Jersey, USA,
5.7 × 3.6 × 1.9 cm.

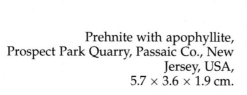

PREHNITE PSEUDOMORPH AFTER ANHYDRITE

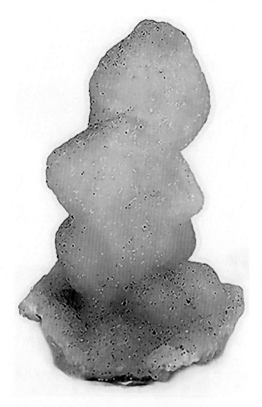

Prehnite pseudomorph after anhydrite,
Upper New Street Quarry, Paterson, New
Jersey, USA, 3.7 × 3.1 × 2.7 cm.

Anhydrite,
Naica, Chihuahua, Mexico,
5.1 × 3.9 × 2.6 cm.

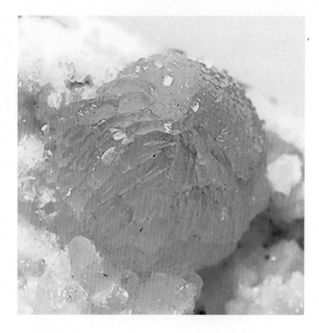

Prehnite on quartz,
Brandberg, Namibia,
9 × 5.4 × 2.5 cm.

PREHNITE PSEUDOMORPH AFTER GLAUBERITE

Glauberite,
Camp Verde, Yavapai County, Arizona,
5.2 × 2.8 × 2.7 cm.

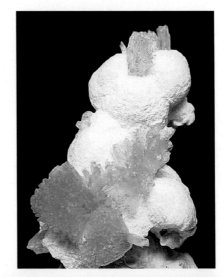

Prehnite pseudomorph after glauberite,
Fanwood Quarry, Union County, New Jersey, USA,
8.4 × 7.2 × 4.7 cm.

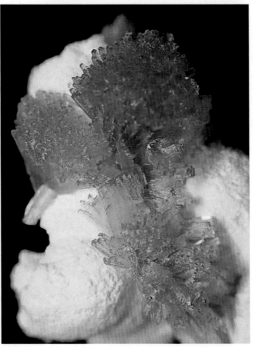

Orange prehnite on xonotlite,
Wessels Mine, Kalahari, South Africa,
3.5 × 2.5 × 2 cm.

PREHNITE PSEUDOMORPH AFTER LAUMONTITE

Prehnite pseudomorph after laumontite, Malad, Mumbai District (Bombay District), Maharashtra, India,
15 × 14 × 12.5 cm.

Laumontite, Himalaya Mine, Mesa Grande, San Diego Co., California, USA,
4.5 × 3 × 2.3 cm.

Prehnite,
Merelani Hills, near Arusha, Tanzania.,
2.1 × 1.7 × 1.5 cm.

CHRYSOCOLLA PSEUDOMORPH AFTER AZURITE

Chrysocolla pseudomorph after malachite after azurite with wulfenite. This is a double pseudomorph showing a change of chemistry twice over to end up with a complete chrysocolla replacement of the original azurite crystal habit, with malachite as a transitional stage,
Whim Creek Copper Mine, Whim Creek, Roebourne Shire,
Western Australia,
Australia,
5.1 × 4.8 × 2.8 cm.

Azurite,
Shi Lu Coppper Mine, Guangdong Province, China,
4 × 3 × 2.75 cm.

Chrysocolla with partial coating of malachite, origin: Star of Congo Mine, Kolwezi, Shaba Province, Congo, 11.4 × 7.0 × 6.0 cm.

Primary malachite on dolomite, Tsumeb Mine, Tsumeb, Namibia,
4.9 × 4.1 × 2.8 cm.

17

Tectosilicates

QUARTZ PSEUDOMORPH AFTER ANHYDRITE

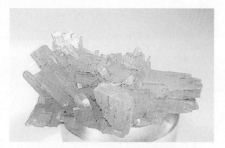

Quartz pseudomorph after anhydrite,
Maricopa, Arizona, USA,
6.4 × 6.4 × 4.5 cm.

Quartz (slightly amethystine in color),
Cerro De La Concordia, Piedra Parada, Vera Cruz, Mexico,
4.6 × 4.4 × 3.2 cm.

Photo Atlas of Mineral Pseudomorphism
http://dx.doi.org/10.1016/B978-0-12-803674-7.00017-7

245

QUARTZ PSEUDOMORPH AFTER BETA-QUARTZ

Quartz paramorph after beta-quartz (beta-quartz is only stable above 572°C and changes to normal (alpha) quartz upon cooling),
Nikolaevskiy Mine, Dalnegorsk, Russia,
4 × 3 × 2.5 cm.

Slightly smoky quartz with purple fluorite and muscovite,
Yaogangxian Mine, Chenzhou, Hunan, China,
14.9 × 8.8 × 4.9 cm.

QUARTZ PSEUDOMORPH AFTER CALCITE

Calcite on sphalerite,
Elmwood Mine, Carthage, Tennessee,
USA,
9.4 × 6 × 6.4 cm.

Quartz pseudomorph after calcite,
Irai, Brazil,
12.1 × 11 × 8.5 cm.

Rose Quartz,
Sapucaia Pegmatite, Governador Valadares,
Minas Gerais, Brazil,
5.4 × 3.4 × 2.3 cm.

QUARTZ PSEUDOMORPH AFTER CALCITE

Quartz pseudomorph after calcite,
Asar Hill, Güğtı, Dursunbey District, Balikesir
Province, Marmara Region, Turkey,
8.9 × 8.5 × 4.5 cm.

Calcite (plumbian) with duftite,
Tsumeb, Namibia,
10 × 8 × 5 cm.

Quartz, slightly smoky in color,
Pederneira Mine, Minas Gerais, Brazil,
7.6 × 4.5 × 4.0 cm.

QUARTZ PSEUDOMORPH AFTER GYPSUM

Quartz pseudomorph after gypsum,
Nebraska, USA,
4.4 × 4.1 × 3.3 cm.

Gypsum,
Norman, Oklahoma, USA,
15.8 × 12.4 × 10.5 cm.

Quartz,
Mogila Mine, Deveti Septemvri Complex,
Madan District, Southern Rhodope
Mountains, Bulgaria,
21.3 × 14.3 × 7.9 cm.

QUARTZ PSEUDOMORPH AFTER DANBURITE

Quartz pseudomorph after danburite,
Charcas, San Luis Potosi, Mexico,
$10 \times 9 \times 8$ cm.

Danburite,
Charcas, San Luis Potosi, Mexico,
$4.4 \times 2.9 \times 2.0$ cm.

Quartz with chlorite inclusion,
Haramosh Mountains, Gilgit, Pakistan,
$6 \times 5 \times 4$ cm.

QUARTZ PSEUDOMORPH AFTER FLUORITE

Quartz pseudomorph after fluorite,
Rock Candy Mine, British Columbia, Canada,
10.6 × 9.8 × 9 cm.

Fluorite with pyrite,
Huanzala Mine, Huanuco Department, Peru,
4.5 × 3.5 × 3 cm.

Quartz,
Bor Pit, Dalnegorsk, Russia,
6.3 × 6 × 5.5 cm.

QUARTZ PSEUDOMORPH AFTER FLUORITE

Quartz pseudomorph after fluorite,
Dalnegorsk, Russia,
6.9 × 6.2 × 5 cm.

Fluorite,
De An Mine, Jian Jiang, Jiangxi Province,
China,
2.1 × 1.7 × 1.1 cm.

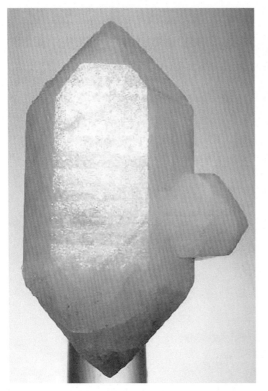

Double terminated quartz,
Auf dem Stein Quarry, Suttrop, Sauerland,
North Rhine-Westphalia, Germany,
2.9 × 1.9 × 1.3 cm.

QUARTZ PSEUDOMORPH AFTER WULFENITE

Quartz pseudomorph after wulfenite,
Finch Mine, Hayden, Banner District, Gila
Co., Arizona, USA,
2.2 × 2.0 × 1.7 cm.

Wulfenite with mimetite,
San Francisco Mine, Sonora, Mexico,
3.5 × 2.75 cm.

Quartz with inverse amethystine phantom,
Brandberg, Namibia,
5.5 × 2.25 × 1.75 cm.

QUARTZ PSEUDOMORPH AFTER TALC

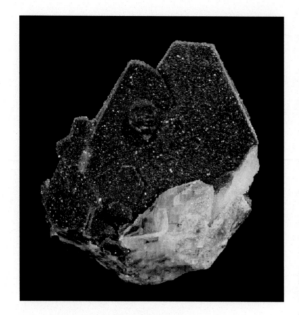

Quartz pseudomorph after talc,
Bahia, Brazil,
21 mm.

Talc,
Gerendorf, Oberwald, Canton Wallis,
Switzerland,
6.7 × 4.8 × 2.8 cm.

Quartz sceptre,
Banska, Stavnica, Slovak Republic,
3 × 2.1 × .7 cm.

CHALCEDONY PSEUDOMORPH AFTER CALCITE

Calcite with hematite and duftite inclusions,
Tsumeb, Namibia,
4.5 × 3.5 × 2.5 cm.

Chalcedony pseudomorph after calcite on
chalcedony,
Jalgaon, Maharashtra Province, India,
13.6 × 8.7 × 7.0 cm.

Chalcedony,
Rio Grande do Sul, Brazil,
23 × 12 × 3.5 cm.

CHALCEDONY PSEUDOMORPH AFTER ARAGONITE

Chalcedony pseudomorph after aragonite,
Big Bend Area, Brewster Co., Texas, USA,
6.5 × 6.2 × 5.4 cm.

Aragonite twin,
Molina de Aragon, Guadalajara, Castile-La
Mancha, Spain (type locality),
3.3 × 3.1 × 2.4 cm.

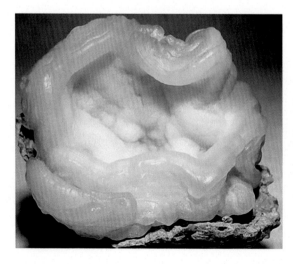

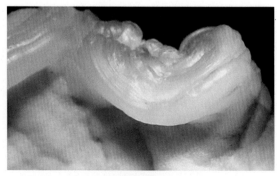

Chalcedony,
Yankee Dog Claim, New Mexico, USA,
Sample size: 6.5 × 4.5 × 5 cm.

CHALCEDONY PSEUDOMORPH AFTER FLUORITE

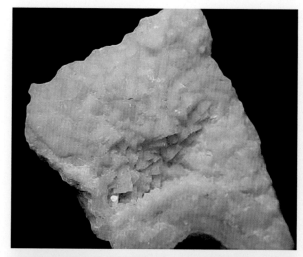

Chalcedony pseudomorph after fluorite,
Trestia, south of Cavnic, Muramures,
Romania,
7 × 5 × 2.5 cm.

Fluorite,
Dalnegorsk, Primorskiy Kray, Russia,
10 × 8.5 × 5.5 cm.

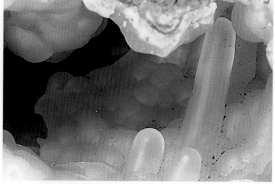

Chalcedony,
Mexico,
13.7 × 10.0 × 9.3 cm.

CHALCEDONY PSEUDOMORPH AFTER QUARTZ

Quartz with Japan law twin,
Siglo XX mine, Llallagua, Bustillo Province,
Potosi Department, Bolivia,
6.5 × 5.3 × 3.8 cm.

Chalcedony pseudomorph after quartz
Japanese twin,
Mkobola District, Mpumalanga Province,
South Africa,
22.0 × 17.0 × 8.5 cm.

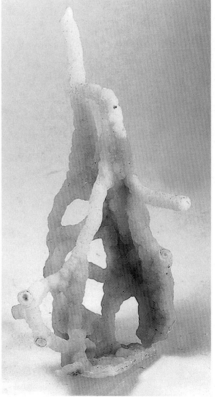

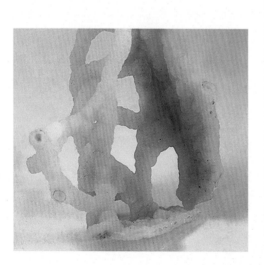

Chalcedony,
Jalgaon, Maharashtra, India,
11.4 × 5.6 × 4.9 cm.

OPAL PSEUDOMORPH AFTER IKAITE

Opal pseudomorph after ikaite. Ikaite is a very rare carbonate that is formed in freezing conditions at or below 0 °C in organic rich, anoxic environments. Just above freezing point the Ikaite dehydrates to calcite (see calcite pseudomorphs),
White Cliffs, New South Wales, Australia,
10 × 9 × 9 cm.

Opal,
Coober-Pedy, South Australia, Australia,
11.0 × 7.2 × 4.8 cm.

OPAL (HYALITE) PSEUDOMORPH AFTER QUARTZ

White hyalite opal pseudomorph after
quartz with spessartine,
Tongbei, Fujian Province, China,
4.4 × 2.4 × 2.2 cm.

Quartz with white phantom,
Saline County, Arkansas, USA,
6.6 × 4.7 × 3.4 cm.

Hyalite opal,
Dalby, Queensland, Australia,
4.0 × 3.2 × 2.4 cm.

AGATE PSEUDOMORPH AFTER BARYTE

Agate pseudomorph after baryte,
nr. Fruita, Mesa County, Colorado, USA,
1.8 × 0.8 × 0.8 cm.

Baryte on quartz pseudomorph after calcite,
Pohla, Erzgebirge, Saxony, Germany,
8.3 × 4.8 × 4.8 cm.

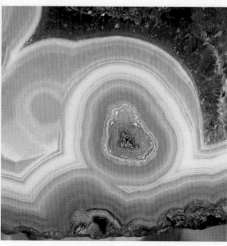

Agate,
Coyamito, Mexico,
8.5 × 6.4 × 4.6 cm.

ORTHOCLASE PSEUDOMORPH AFTER LEUCITE

Orthoclase pseudomorph after leucite, Oberwiesenthal, Germany, in the Erzgebirge, near the border between Bohemia, Czech Republic, and Obersachsen, Germany, 6.4 × 5.7 × 5.5 cm.

Leucite with small augite crystal on the side, Roccamonfina, Caserta Province, Campania, Italy, 2.5 × 2.3 × 1.8 cm.

Spessartine on Baveno twinned orthoclase and smoky quartz, Tongbei, Fujian Province, China, 3.5 × 3.5 × 3.2 cm.

ORTHOCLASE PSEUDOMORPH AFTER SCAPOLITE

Scapolite,
Morogoro, Tanzania,
2.2 × 1 × 0.8 cm.

Orthoclase pseudomorph after scapolite,
Spain Mine, Griffith Twnshp., Renfrew Co.,
Ontario, Canada,
5.6 × 3.4 × 3.3 cm.

Orthoclase,
Montana, USA,
6.6 × 3.7 × 3.4 cm.

LAZURITE PSEUDOMORPH AFTER MUSCOVITE

Lazurite pseudomorph after muscovite,
Koksha Valley, Jurm District, Badakhshan
Province, Afghanistan,
6.1 × 6 × 5 cm.

Muscovite with albite,
Pack Rat Mine, Jacumba, San Diego Co.,
California, USA,
12.0 × 9.0 × 7.5 cm.

Lazurite on marble,
Sar-E-Sang, Badakshan Province, Afghanistan,
4.2 × 3.8 × 2.6 cm.

CANCRINITE PSEUDOMORPH AFTER NATROLITE

Cancrinite pseudomorph after natrolite
with rhodochrosite,
Mont St. Hilaire, Quebec, Canada,
4 × 2 × 1.75 cm.

Natrolite with salmon-coloured serandite
and a small needle of aegirine,
Mont. St-Hilaire, Quebec, Canada,
3.7 × 2 × 1.7 cm.

Cancrinite,
Litchfield, Kennebec Co., Maine, USA,
6.7 × 5.0 × 3.1 cm.

CHABAZITE PSEUDOMORPH AFTER GYPSUM

Chabazite pseudomorph after gypsum,
Daye Co., China,
14 cm.

Gypsum,
Willow Creek, near Nanton, Alberta,
Canada,
$4.9 \times 2.4 \times 1.5$ cm.

Chabazite-Ca,
Wasson''s Bluff, Parrsboro, Nova Scotia,
Canada,
$4.9 \times 3.6 \times 2.1$ cm.

NATROLITE PSEUDOMORPHITE AFTER SERANDITE

Analcime on natrolite pseudomorph after
serandite,
Mt. St-Hilaire, Quebec, Canada,
8.7 × 5.4 × 5.0 cm.

Serandite with analcime,
Mt. St. Hilaire, Quebec, Canada,
4.3 × 3.0 × 2.5 cm.

Natrolite in basalt vug,
Aurangabad, Maharashtra, India,
6.4 × 5.5 × 5.0 cm.

BREWSTERITE PSEUDOMORPH AFTER CALCITE

Brewsterite cast after calcite (pseudomorph)
Yellow Lake, near Ollala, British Columbia,
Canada,
3.7 × 2.2 × 2.1 cm.

Calcite, Tsumeb, Namibia,
9.5 × 3 × 2 cm (top), 6 × 5 × 4 cm.

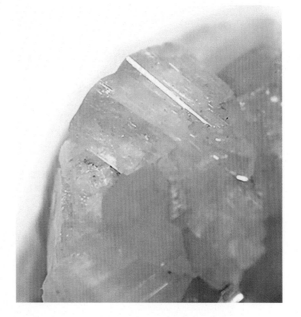

Brewsterite,
Yellow Lake, near Ollala, British Columbia, Canada,
2.9 × 2.5 × 1.6 cm.

Fossils

RHODOCHROSITE PSEUDOMORPH AFTER SHELL

Rhodochrostite pseudomorph after shell,
Kerch, Ukraine,
3.1 × 3 × 3 cm.

Rhodochrosite with quartz and fluorite,
Strawberry Pocket, Sweet Home Mine, Alma, Colorado, USA,
6.5 × 6 × 3 cm.

Photo Atlas of Mineral Pseudomorphism
http://dx.doi.org/10.1016/B978-0-12-803674-7.00018-9

RHODOCHROSITE PSEUDOMORPH AFTER SNAIL

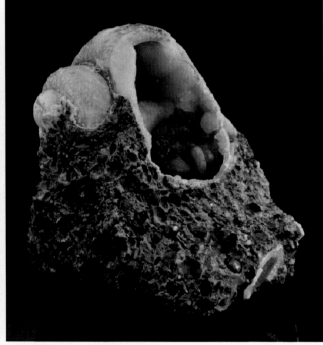

Rhodochrosite pseudomorph after snail,
Kerch, Ukraine,
4.2 × 4 × 4 cm.

Rhodochrosite with quartz, hubnerite, and fluorite,
Clay Pocket, Sweet Home Mine, Alma, Colorado, USA,
9.5 × 6 × 5 cm.

APATITE PSEUDOMORPH AFTER WHALE BONE

Apatite pseudomorph after whale bone,
Kerch, Ukraine,
13.2 × 6.5 × 6.5 mm.

Fluorapatite with beryl, variety aquamarine
(blue), on muscovite,
Hunza Valley, Nagar, Pakistan,
8.6 × 6.7 × 4.8 cm.

TURQUOISE PSEUDOMORPH AFTER MARMOT JAW

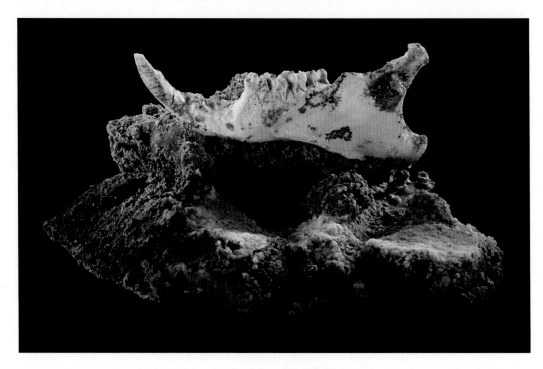

Turquoise pseudomorph after marmot jaw,
Potosi Mountain area, Clark Co., Nevada, USA,
Lavinsky collection,
10 × 7.8 × 6 cm.

Turquoise,
Bishop Mine, Lynch Station, Campbell Co.,
Virginia, USA,
4.5 × 2.2 × 1.6 cm.

VIVIANITE PSEUDOMORPH AFTER SHELL

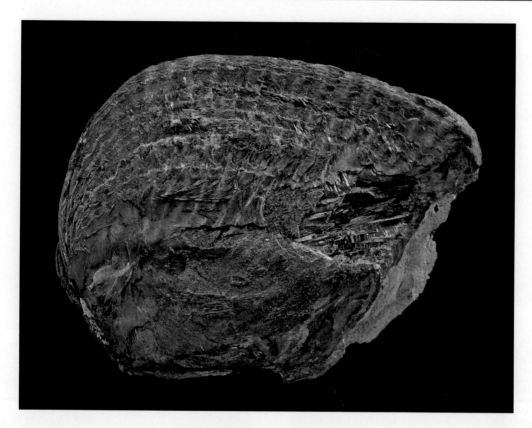

Vivianite pseudomorph after shell,
Kerch, Ukraine,
$3 \times 3 \times 2.5$ cm.

Vivianite,
Tomokoni adit, near Canutillos, Colavi, Potosi Department, Bolivia,
$4.2 \times 2.7 \times 2.3$ cm.

AGATE PSEUDOMORPH AFTER SNAIL

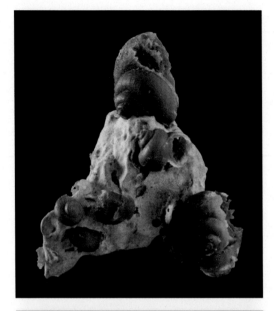

Agate pseudomorph after snails
Roadcut, Mesa Co., Colorado, U.S.A.
3.8 × 3.8 × 3.5 cm

Agate (polished cut),
Rio Grande do Sul, Brazil,
6 × 6 × 4 cm.

OPAL PSEUDOMORPH AFTER AMMONITE

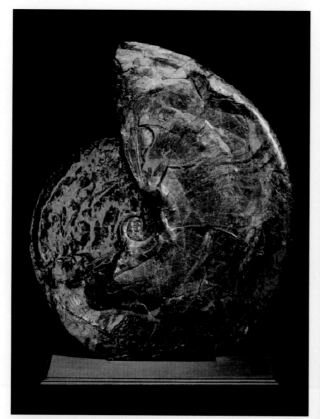

Opal pseudomorph after ammonite,
Calgary, Canada,
Lavinsky collection,
54 cm tall, 48 cm wide, up to 4 cm deep.

Opal,
Quilpie, Queensland, Australia,
11.3 × 6.9 × 5.5 cm.

Index

Printed in the United States
By Bookmasters